Recent Ti[redacted]

486 **Brian D. [redacted]**-Shelton theory and Vogan's problem for generalized principal series, 1993

485 **Paul Feit,** Axiomization of passage from "local" structure to "global" object, 1993

484 **Takehiko Yamanouchi,** Duality for actions and coactions of measured groupoids on von Neumann algebras, 1993

483 **Patrick Fitzpatrick and Jacobo Pejsachowicz,** Orientation and the Leray-Schauder theory for fully nonlinear elliptic boundary value problems, 1993

482 **Robert Gordon,** G-categories, 1993

481 **Jorge Ize, Ivar Massabo, and Alfonso Vignoli,** Degree theory for equivariant maps, the general S^1-action, 1992

480 **L. Š. Grinblat,** On sets not belonging to algebras of subsets, 1992

479 **Percy Deift, Luen-Chau Li, and Carlos Tomei,** Loop groups, discrete versions of some classical integrable systems, and rank 2 extensions, 1992

478 **Henry C. Wente,** Constant mean curvature immersions of Enneper type, 1992

477 **George E. Andrews, Bruce C. Berndt, Lisa Jacobsen, and Robert L. Lamphere,** The continued fractions found in the unorganized portions of Ramanujan's notebooks, 1992

476 **Thomas C. Hales,** The subregular germ of orbital integrals, 1992

475 **Kazuaki Taira,** On the existence of Feller semigroups with boundary conditions, 1992

474 **Francisco González-Acuña and Wilbur C. Whitten,** Imbeddings of three-manifold groups, 1992

473 **Ian Anderson and Gerard Thompson,** The inverse problem of the calculus of variations for ordinary differential equations, 1992

472 **Stephen W. Semmes,** A generalization of riemann mappings and geometric structures on a space of domains in $\mathbf{C}^n$, 1992

471 **Michael L. Mihalik and Steven T. Tschantz,** Semistability of amalgamated products and HNN-extensions, 1992

470 **Daniel K. Nakano,** Projective modules over Lie algebras of Cartan type, 1992

469 **Dennis A. Hejhal,** Eigenvalues of the Laplacian for Hecke triangle groups, 1992

468 **Roger Kraft,** Intersections of thick Cantor sets, 1992

467 **Randolph James Schilling,** Neumann systems for the algebraic AKNS problem, 1992

466 **Shari A. Prevost,** Vertex algebras and integral bases for the enveloping algebras of affine Lie algebras, 1992

465 **Steven Zelditch,** Selberg trace formulae and equidistribution theorems for closed geodesics and Laplace eigenfunctions: finite area surfaces, 1992

464 **John Fay,** Kernel functions, analytic torsion, and moduli spaces, 1992

463 **Bruce Reznick,** Sums of even powers of real linear forms, 1992

462 **Toshiyuki Kobayashi,** Singular unitary representations and discrete series for indefinite Stiefel manifolds $U(p,q;\mathbb{F})/U(p-m,q;\mathbb{F})$, 1992

461 **Andrew Kustin and Bernd Ulrich,** A family of complexes associated to an almost alternating map, with application to residual intersections, 1992

460 **Victor Reiner,** Quotients of coxeter complexes and P-partitions, 1992

459 **Jonathan Arazy and Yaakov Friedman,** Contractive projections in C_p, 1992

458 **Charles A. Akemann and Joel Anderson,** Lyapunov theorems for operator algebras, 1991

457 **Norihiko Minami,** Multiplicative homology operations and transfer, 1991

456 **Michał Misiurewicz and Zbigniew Nitecki,** Combinatorial patterns for maps of the interval, 1991

455 **Mark G. Davidson, Thomas J. Enright and Ronald J. Stanke,** Differential operators and highest weight representations, 1991

(*Continued in the back of this publication*)

[redacted]

Enright-Shelton Theory and Vogan's Problem for Generalized Principal Series

MEMOIRS
of the
American Mathematical Society

Number 486

Enright-Shelton Theory and Vogan's Problem for Generalized Principal Series

Brian D. Boe
David H. Collingwood

March 1993 • Volume 102 • Number 486 (first of 4 numbers) • ISSN 0065-9266

American Mathematical Society
Providence, Rhode Island

1991 *Mathematics Subject Classification.*
Primary 22E46, 22E47.

Library of Congress Cataloging-in-Publication Data

Boe, Brian D., 1956–
Enright-Shelton theory and Vogan's problem for generalized principal series/Brian D. Boe, David H. Collingwood.
p. cm. – (Memoirs of the American Mathematical Society, ISSN 0065-9266; no. 486)
Includes bibliographical references.
ISBN 0-8218-2547-X
1. Semisimple Lie groups. 2. Representations of groups. I. Collingwood, David H. II. Title. III. Series.
QA3.A57 no. 486
[QA387]
510 s–dc20 92-38215
[512′.55] CIP

Memoirs of the American Mathematical Society

This journal is devoted entirely to research in pure and applied mathematics.

Subscription information. The 1993 subscription begins with Number 482 and consists of six mailings, each containing one or more numbers. Subscription prices for 1993 are $336 list, $269 institutional member. A late charge of 10% of the subscription price will be imposed on orders received from nonmembers after January 1 of the subscription year. Subscribers outside the United States and India must pay a postage surcharge of $25; subscribers in India must pay a postage surcharge of $43. Expedited delivery to destinations in North America $30; elsewhere $92. Each number may be ordered separately; *please specify number* when ordering an individual number. For prices and titles of recently released numbers, see the New Publications sections of the *Notices of the American Mathematical Society.*

Back number information. For back issues see the *AMS Catalog of Publications.*

Subscriptions and orders should be addressed to the American Mathematical Society, P. O. Box 1571, Annex Station, Providence, RI 02901-1571. *All orders must be accompanied by payment.* Other correspondence should be addressed to Box 6248, Providence, RI 02940-6248.

Memoirs of the American Mathematical Society is published bimonthly (each volume consisting usually of more than one number) by the American Mathematical Society at 201 Charles Street, Providence, RI 02904-2213. Second-class postage paid at Providence, Rhode Island. Postmaster: Send address changes to Memoirs, American Mathematical Society, P. O. Box 6248, Providence, RI 02940-6248.

Printed in the United States of America.
This volume was printed directly from author-prepared copy.
The paper used in this book is acid-free and falls within the guidelines
established to ensure permanence and durability. ♾
10 9 8 7 6 5 4 3 2 1 98 97 96 95 94 93

CONTENTS

ABSTRACT

We study the composition series of generalized principal series representations induced from a maximal cuspidal parabolic subgroup of a real reductive Lie group. In the Hermitian symmetric setting, we investigate when such induced representations are multiplicity free and the problem of describing their composition factors in closed form. The closed formulas obtained are very much in the spirit of Enright-Shelton's work for highest weight modules. Our results shed new light on a problem posed by D. Vogan in 1981 and provide counterexamples to an old conjecture of Speh-Vogan.

Key words and phrases. representations of semisimple Lie groups, principal series representations, Hermitian symmetric, composition series, standard modules, Kazhdan-Lusztig conjectures.

1. INTRODUCTION

Let G be a connected semisimple real matrix group and $P \subset G$ a cuspidal parabolic subgroup. Fix a Langlands decomposition $P=MAN$ of P, with N the unipotent radical and A a vector group. Let σ be a discrete series representation of M and ν a (non-unitary) character of A. We call the induced Harish-Chandra module $\mathrm{Ind}_P^G(\sigma \otimes \nu \otimes 1)$ (normalized induction) a *P-generalized principal series representation.* Denote by $\mathcal{F}_P$ the class of all P-generalized principal series representations with regular integral infinitesimal character. In turn, $\mathcal{F}_{\max}$ will be the union of all $\mathcal{F}_P$, where P ranges over the maximal cuspidal parabolic subgroups of G. The representations in $\mathcal{F}_{\max}$ play a fundamental role in the harmonic analysis and representation theory of G, which strongly motivates a detailed analysis of their structure and properties. To begin with, the *Speh-Vogan Theorem* [20] gives necessary and sufficient conditions for the reduciblity of such representations and the *Kazhdan-Lusztig Conjecture* (proved in [23]) provides an algorithm to compute their irreducible composition factors. However, due to the highly recursive nature of this algorithm, it is often quite hard to *a priori* extract qualitative features of the generalized principal series representations. The starting point for our work is the following problem, posed by D.Vogan in 1981 (modeled on an earlier conjecture of Speh-Vogan [20, (6.20)]).

(1.1) *Problem 3 ([21, p.737]): Let $P=MAN$ be a cuspidal parabolic subgroup with $\dim_{\mathbb{R}} A=1$. Describe the composition factors of each P-generalized principal series representation in $\mathcal{F}_P$. Do the composition factors always occur with multiplicity one?*

At the time of its proposal, Vogan was aware of this problem's solution in the case when G has real rank one; in that case, multiplicity one holds and it is a fairly easy matter to write out closed formulas for the composition series (in fact, for any *Loewy series*), see [7]. However, beyond the real rank one setting, the situation rapidly complicates; a convincing example is the real rank two Hermitian symmetric form of E_6 (discussed in chapter 7) which admits 513 distinct irreducible Harish-Chandra modules with a fixed regular integral infinitesimal character.

Received by the editors 23 January 1991.
The first author was supported in part by a University of Georgia Faculty Research Grant.
The second author was supported in part by NSA grant MDA904-90-H-4041 and the University of Washington Graduate Research Fund

In this paper, we will restrict our attention to the case when *(G,K)* **is an irreducible Hermitian symmetric pair of non-compact type**; a list of all such G, up to local isomorphism, is recalled in chapter 2. We present a series of results related to (1.1), from which a picture will emerge which is strikingly similar to our current state of understanding of categories of highest weight representations of *holomorphic type*. Because of these similarities, let's review some of the basic features common to such highest weight module categories. First, recall that each complex structure on the Hermitian symmetric space G/K gives rise to a family of irreducible highest weight representations of G; this is reviewed at the beginning of chapter 4. These highest weight representations arise as irreducible quotients of *generalized Verma modules* algebraically induced from a complex parabolic subalgebra $\mathfrak{q}^\sharp = \mathfrak{k} \oplus \mathfrak{u}$ of $\mathfrak{g}$, where $\mathfrak{k}$ is the complexified Lie algebra of K and $\mathfrak{q}^\sharp$ is determined by the compact dual symmetric space $(G/K)\check{} = \mathbf{G}/\mathbf{Q}^\sharp$, $\mathbf{Q}^\sharp$ a parabolic subgroup of the complexification $\mathbf{G}$ of G, see [14]. If we form the category $\mathcal{O}(\mathfrak{g}, \mathfrak{q}^\sharp)$ of all finitely generated $\mathfrak{U}(\mathfrak{g})$-modules which are $\mathfrak{q}^\sharp$-locally finite, $\mathfrak{k}$-semisimple and have regular integral generalized infinitesimal character, we obtain a variant of the classical *Bernstein-Gel'fand-Gel'fand category* $\mathcal{O}$. During the 1980's, through the pioneering work of Enright and Shelton, our understanding of the categories $\mathcal{O}(\mathfrak{g}, \mathfrak{q}^\sharp)$ dramatically improved, leading to the following qualitative results [10], [11]:

(1.2a) The generalized Verma modules in $\mathcal{O}(\mathfrak{g}, \mathfrak{q}^\sharp)$ are multiplicity free;

(b) There exist closed formulas for the radical series of each generalized Verma module and all Loewy series coincide. This is explained further in (1.3) below;

(c) There exists a linear function $p_\mathfrak{g}(r_G)$, in the real rank r_G of G, which bounds the Loewy length of every generalized Verma module in $\mathcal{O}(\mathfrak{g}, \mathfrak{q}^\sharp)$.

If we were to replace the families $\mathcal{F}_P$ in (1.1) by the categories $\mathcal{O}(\mathfrak{g}, \mathfrak{q}^\sharp)$, then we see that (1.2) is a very satisfactory answer to the obvious analog of Vogan's problem in the highest weight module setting. With this in mind, we take (1.2) as a model for our theorems on P-generalized principal series representations. Our results are of three types: multiplicity one theorems (Theorems A, B, D, E); combinatorial closed formulas for a Loewy series (Theorems 4.8, E, 7.11); Loewy length estimates (Theorems G, H).

Our first results will analyze necessary and sufficient conditions for the families $\mathcal{F}_{\max}$ to be *multiplicity free*, meaning that every induced module in $\mathcal{F}_{\max}$ has a multiplicity free composition series. From these results the rather subtle nature of (1.1) for higher real rank groups will become apparent.

THEOREM A. *Assume G is simple of Hermitian symmetric type and not locally isomorphic to $Sp_n\mathbb{R}$ or $SO^*(2n)$. Then $\mathcal{F}_{\max}$ is multiplicity free.*

THEOREM B. *Assume $G = Sp_n\mathbb{R}$. Then $\mathcal{F}_{\max}$ is multiplicity free if and only if $n \leq 3$.*

As a consequence, Theorem B provides us with a counterexample (indeed, a family of such) to an old conjecture of Speh-Vogan [20, (6.20)]. In the case $G = SO^*(2n)$, the situation is not as yet completely understood. Computer calculations have verified that $\mathcal{F}_{\max}$ is multiplicity free, whenever $n \leq 7$. For general n, we have a partial result, as described in Corollary 6.22; roughly speaking, this result says that at least "half" of the modules in $\mathcal{F}_{\max}$ are multiplicity free. In view of Theorems A and B, showing that the remaining modules in $\mathcal{F}_{\max}$ are multiplicity free would prove the following conjecture.

CONJECTURE C. *Assume G is simple of Hermitian symmetric type. Then the following are equivalent:*

(i) $\mathcal{F}_{\max}$ is multiplicity free;

(ii) G is not locally isomorphic to $Sp_n\mathbb{R}, n \geq 4$.

Our proof of Theorem A will follow from "Enright-Shelton type" closed formulas for a Loewy series of each module in $\mathcal{F}_{\max}$, as described below.

When studying the structure and properties of the induced modules $\mathrm{Ind}_P^G(\sigma \otimes \nu \otimes 1)$, it is natural to investigate the extent to which their structural complexity is controlled by the inducing data. Intuitively, the structure should be simplest when P is a maximal cuspidal parabolic subgroup and σ is a "simple" discrete series representation for M. Among the classes of discrete series representations for M, the holomorphic discrete series, if such exist, are the most elementary. Our intuition is reinforced by our final multiplicity one theorem.

THEOREM D. *Let G be simple of Hermitian type and $P = MAN$ a maximal cuspidal parabolic subgroup of G where the noncompact semisimple factor of M is both simple and of Hermitian type. If σ is a holomorphic discrete series representation of M, then $\mathrm{Ind}_P^G(\sigma \otimes \nu \otimes 1)$ is multiplicity free.*

We introduce the category $\mathcal{HC}_o$ of Harish-Chandra modules for G with the same infinitesimal character as some fixed finite dimensional representation of G. Referring to chapter 2 for unexplained notation and terminology, we have the basis of irreducible representations $\{\, \overline{\pi}(\delta) \mid \delta \in \mathcal{D} \,\}$ and the basis of standard modules $\{\, \pi(\delta) \mid \delta \in \mathcal{D} \,\}$ for the Grothendieck group of virtual characters of $\mathcal{HC}_o$. Observe, if P is a maximal cuspidal parabolic subgroup of G, then a solution to (1.1) will follow if we can describe a Loewy filtration for each standard module $\pi(\delta)$ whose underlying Langlands data involves P. We will often want to single out the elements of the parameter set $\mathcal{D}$ with the property that the associated standard module is induced from a maximal cuspidal parabolic subgroup of G; such parameters define a subset denoted $\mathcal{D}^{\max} \subset \mathcal{D}$.

When studying questions related to characters and radical filtrations of standard modules or generalized Verma modules, it is common practice to work within the framework of the *Hecke modules* $\mathcal{M}_{\mathbf{K}}$ and $\mathcal{M}_{\mathbf{Q}^\sharp}$ associated to the parameter sets $\mathcal{D}$ and $W^{\mathbf{Q}^\sharp}$ (see chapter 2). The modules are free on these parameter sets as $\mathbb{Z}[u^{1/2}, u^{-1/2}]$-modules, but additionally carry an action of the Hecke algebra associated to the underlying Weyl group; this action arises as a u-deformation of the coherent continuation action on characters. In addition to the standard bases $\mathcal{D}$ and $W^{\mathbf{Q}^\sharp}$ for these modules, there exist *self-dual* bases $\{\, \hat{C}_\delta \mid \delta \in \mathcal{D} \,\}$ and $\{\, \hat{C}_w \mid w \in W^{\mathbf{Q}^\sharp} \,\}$ which arise through the intersection cohomology complexes attached to $\mathbf{K}$- and $\mathbf{Q}^\sharp$-orbit closures in the flag variety. As reviewed in chapter 2, giving formulas for a standard parameter in $\mathcal{D}$ or $W^{\mathbf{Q}^\sharp}$ in this self-dual basis is tantamount to a description of the radical filtration of the corresponding standard module or generalized Verma module.

Our combinatorial formulas are motivated by the work of Enright and Shelton in [11, Part II]. Because this motivation is so fundamental to our paper, let's recall the basic flavor of the Enright-Shelton theory. Referring to the sequel for any undefined notation, Enright and Shelton associate to each $w \in W^{\mathbf{Q}^\sharp}$ a collection $\mathcal{E}_w$ of sets of orthogonal roots. Given $\Omega \in \mathcal{E}_w$, they define a new element $\overline{wr_\Omega} \in W^{\mathbf{Q}^\sharp}$. Define a map $\mathbb{E}_w$ from the collection $\mathcal{E}_w$ into the Hecke module $\mathcal{M}_{\mathbf{Q}^\sharp}$, by $\mathbb{E}_w(\Omega) = \hat{C}_{\overline{wr_\Omega}}$. The main result of [10] establishes the Hecke module formula

$$w = \sum_{\Omega \in \mathcal{E}_w} u^{-|\Omega^+|/2}\, \mathbb{E}_w(\Omega). \tag{1.3}$$

From (1.3), one can read off a weight filtration for the corresponding generalized Verma module N_w, which coincides with the radical filtration. In particular, we find that the irreducible highest weight module L_y occurs exactly once as a composition factor of N_w, and in the ith-layer of the radical filtration of N_w, if and only if the coefficient of $\hat{C}_y$ in (1.3) is precisely $u^{-i/2}$.

To generalize the Enright-Shelton program to our Harish-Chandra module setting, we adopt the following philosophy:

(1.4) *To each $\delta \in \mathcal{D}^{\max}$, associate a collection $\mathcal{E}_\delta$ of sets Ω of roots. Describe a non-recursive algorithm $\mathbb{E}_\delta$ which associates a composition factor (or factors) $\overline{\pi}(\gamma_\Omega)$ of $\pi(\delta)$ to each $\Omega \in \mathcal{E}_\delta$.*

Referring the reader to chapter 2 for undefined terminology, Theorem 4.8, Theorem E and Theorem 7.11 will each describe a map $\mathbb{E}_\delta : \mathcal{E}_\delta \longmapsto \mathcal{M}_{\mathbf{K}}$, which attaches a Hecke module element $\mathbb{E}_\delta(\Omega) \in \mathcal{M}_{\mathbf{K}}$, expressed in the self-dual basis arising from intersection cohomology complexes of $\mathbf{K}$-orbit closures. Notationally, we will have closed formulas for expressions of the form

$$\mathbb{E}_\delta(\Omega) = \sum_{\gamma \in \mathcal{D}} E_{\gamma,\delta,\Omega}(u)\widehat{C_\gamma}, \quad E_{\gamma,\delta,\Omega}(u) \in \mathbb{Z}[u^{1/2}, u^{-1/2}].$$

Depending on the hypothesis, these three theorems will then assert:

The irreducible module $\overline{\pi}(\gamma)$ occurs precisely k_γ times in the ith-layer of the radical filtration of $\pi(\delta)$ if and only if $\pm k_\gamma$ is the coefficient of $u^{-i/2}\widehat{C_\gamma}$ in

$$\sum_{\Omega \in \mathcal{E}_\delta} u^{-|\Omega^+|/2}\mathbb{E}_\delta(\Omega). \tag{1.5}$$

In fact, these irreducible composition factors occur with multiplicity one and every composition factor arises in this way. In other words, in $\mathcal{M}_{\mathbf{K}}$, $\delta = \sum_{\Omega \in \mathcal{E}_\delta} u^{-|\Omega^+|/2}\mathbb{E}_\delta(\Omega)$.[1]

[1] Depending on our conventions, the Hecke modules we use may introduce expressions (1.5) involving "minus signs."

Theorem E in chapter 6 deals with the case $G = SU(p,q)$ and involves a generalization of the Enright-Shelton $\mathcal{E}_w$ construction. We maintain the idea of associating (almost always) a unique irreducible composition factor $\overline{\pi}(\gamma_\Omega)$ of $\pi(\delta)$ to each $\Omega \in \mathcal{E}_\delta$, but, we pay a price in that the roots in the sets $\Omega \in \mathcal{E}_\delta$ need not be mutually orthogonal. Here is a formulation of Theorem E, in the spirit of (1.3)–(1.5).

THEOREM E. *Let $G = SU(p,q)$ and assume $\delta \in \mathcal{D}^{\max}$. For each $\Omega \in \mathcal{E}_\delta$, let $\mathbb{E}_\delta(\Omega) = \hat{C}_{\delta\star r_{\Omega^+}}$, as defined in chapter 6. Then $\overline{\pi}(\gamma)$ is a composition factor of $\pi(\delta)$ if and only if $\hat{C}_\gamma$ occurs in $\mathbb{E}_\delta(\Omega)$, for some $\Omega \in \mathcal{E}_\delta$. All composition factors of $\pi(\delta)$ occur with multiplicity one. The composition factor(s) $\overline{\pi}(\gamma)$ attached to $\Omega \in \mathcal{E}_\delta$ lie(s) in the $|\Omega^+|$-layer of the radical filtration of $\pi(\delta)$.*

The main result of chapter 4 is Theorem 4.8, which applies when $\pi(\delta)$ is induced from a holomorphic discrete series. Here, the collection of sets $\mathcal{E}_\delta$ will coincide with an Enright-Shelton collection $\mathcal{E}_w$ for some appropriate w attached to δ, and the map $\mathbb{E}_\delta$ depends on the orbit decomposition result in [9]. The use of the Enright-Shelton set $\mathcal{E}_w$ will necessitate that we typically associate two irreducible composition factors of $\pi(\delta)$ to each $\Omega \in \mathcal{E}_\delta$. As a corollary, we describe the radical filtration of any $\pi(\delta) \in \mathcal{F}_{\max}$, where $\overline{\pi}(\delta)$ is a highest weight module. This result can then be interpreted as the natural analog of Enright-Shelton's results [11] in the category $\mathcal{HC}_o$ setting. In the process of proving Theorem D and Theorem 4.8, we will isolate some interesting geometry exhibited by certain stratifications of the $\mathbf{Q}^\sharp$-orbits; cf. Lemma 4.12. In fact, we believe that appropriately interpreted, these special properties explain the close analogy and subtle differences between our results and the results in [10] and [11].

Theorem 7.11 is in the same spirit as Theorem 4.8 and applies to any module in $\mathcal{F}_{\max}$, in the exceptional cases. This result implies

THEOREM F. *Let G be of exceptional Hermitian symmetric type and $\delta \in \mathcal{D}^{\max}$. Let $\mathcal{O}_w$ be the $\mathbf{Q}^\sharp$-orbit containing the $\mathbf{K}$-orbit determined by δ. Then $\overline{\pi}(\gamma)$ is a composition factor of $\pi(\delta)$ only if the $\mathbf{K}$-orbit determined by γ lies in the $\mathbf{Q}^\sharp$-orbit parametrized by $\overline{wr_\Omega}$ for some $\Omega \in \mathcal{E}_w$.*

This Theorem suggests the general conjecture that the irreducible composition factors of $\pi(\delta)$ are determined "up to $\mathbf{Q}^\sharp$-orbit" by the Enright-Shelton theory of category $\mathcal{O}(\mathfrak{g}, \mathfrak{q}^\sharp)$; see Conjecture 7.2.

A non-standard concept, which is absolutely central to our work, is the notion of a *pyramid* or *G-pyramid* inside $\mathcal{D}$; these ideas are discussed in (2.4). The rough idea is that the set $\mathcal{D}^{\max}$ can be decomposed into a union of posets $\mathcal{D}_P^{\max}$, and

$$\mathcal{D}_P^{\max} = \bigcup_{\sigma \in M_{\mathrm{ds}}} \mathcal{P}_{\sigma,M}, \tag{1.6}$$

where M_{ds} indexes the equivalence classes of discrete series on M with a fixed regular integral infinitesimal character. Each $\mathcal{P}_{\sigma,M}$, called a *pyramid*, will look like the "upper half" of a Bruhat poset $W^{\mathbf{P}}$, where the associated category $\mathcal{O}(\mathfrak{g}, \mathfrak{p})$ is of multiplicity free type as studied in [2] and [7]. Given a pyramid

$\mathcal{P}_{\sigma,M}$, we can consider the set $\widehat{\mathcal{P}}_{\sigma,M}$ of all elements in $\mathcal{D}$ dominated by some element of $\mathcal{P}_{\sigma,M}$ under the G-order on $\mathcal{D}$; we refer to $\widehat{\mathcal{P}}_{\sigma,M}$ as a *G-pyramid*. Whereas the pyramids parametrize the generalized principal series of interest in (1.1), the G-pyramids control the composition factors of such. It is well worth isolating the following remark:

(1.7) Let $\delta \in \mathcal{D}^{\max}$, then the source of all our difficulties stems from the elementary observation that $\overline{\pi}(\gamma)$ may be a composition factor of $\pi(\delta)$, with $\gamma \notin \mathcal{D}^{\max}$.

If $P = MAN$, then the number of pyramids needed in the decomposition (1.6) will be determined by the number of equivalence classes of discrete series representations for M having a fixed infinitesimal character. In all cases except $SO_e(2, 2n-1)$, M will also be of Hermitian type and one of the pyramids will correspond to P-generalized principal series induced from holomorphic discrete series. The G-pyramid attached to this holomorphic induction will be manageable, corresponding to our ability in Theorem 4.8 and Theorem D to reduce to the Enright-Shelton category $\mathcal{O}(\mathfrak{g}, \mathfrak{q}^\sharp)$ considerations. By contrast, as the inducing data moves away from holomorphic type (say toward a large discrete series representation for something like $SU(p,p)$) the G-pyramids become extremely complicated, leading to a corresponding complexity in the radical filtration of the so parametrized $\pi(\delta)$.

Our final two results deal with the Loewy length of a standard module attached to $\delta \in \mathcal{D}^{\max}$. As discussed at the beginning of chapter 8, we have an easily computed theoretical maximum Loewy length for any standard module attached to a parameter in $\mathcal{D}^{\max}$, denoted $b_{\mathbf{K}}$ and computed in Table 8.1. Let $\ell\ell_{\mathbf{K},\max}$ denote the maximum Loewy length acheived by a standard module attached to a parameter in $\mathcal{D}^{\max}$. (It should be noted that the definitions of $b_{\mathbf{K}}$ and $\ell\ell_{\mathbf{K},\max}$ are unchanged if we instead vary over modules in $\mathcal{F}_{\max}$.) Our first result gives a necessary and sufficient condition for the theoretical maximum and actual bounds to coincide.

THEOREM G. *Assume G is simple of Hermitian type and $\mathcal{F}_{\max}$ is multiplicity free. Then $b_{\mathbf{K}} = \ell\ell_{\mathbf{K},\max}$ if and only if G is quasi-split.*

THEOREM H. *Let G be simple of Hermitian symmetric type. Then there exists a linear function $p_G(r_G)$, in the real rank r_G of G, so that $\ell\ell(\pi) \le p_G(r_G)$, for all $\pi \in \mathcal{F}_{\max}$.*

The linear functions of Theorem H, together with related data and the analogous information for the highest weight categories $\mathcal{O}(\mathfrak{g}, \mathfrak{q}^\sharp)$ are tabulated in Table 8.1.

We conclude this introduction with a discussion of the outline of the paper and the strategy of proof. As already noted, chapter 2 contains notation and terminology used throughout the sequel. For the most part we use standard conventions in the current literature. In chapter 3, we prove Theorem B. This is of interest, because it offers a counterexample to the multiplicity one question in (1.1). The proofs involve explicit data for the group $Sp_n\mathbb{R}$, $n = 3, 4$; the case of $Sp_2\mathbb{R}$ appears in [22].

In chapter 4, we prove Theorem D and Theorem 4.8, which depend upon understanding the decomposition of each $\mathbf{Q}^\sharp$-orbit into $\mathbf{K}$-orbits, as described in [9]. In turn, this depends on the recent work of Richardson-Springer [18]. A solution to (1.1) in the case $G = SO_e(2, N)$ is given in chapter 5. This nearly follows from Theorem 4.8, except that we need to handle the situation in $SO_e(2, 2n-1)$ where P is a maximal cuspidal parabolic subgroup whose Levi factor is not of Hermitian type; the simple part of its Levi factor is locally isomorphic to $SO(1, 2n-2)$.

In chapter 6, we work with $SU(p, q)$ and prove Theorem E. An important point here is that we actually prove a stronger version of the theorem, which holds for any $\delta \in \widehat{\mathcal{P}}_{\sigma,M}$. Here, the similarities with the ideas of Enright-Shelton should be striking. We obtain a consequence for the case of $SO^*(2n)$, which motivates our multiplicity one conjecture. The chapter ends with an example of Theorem E for the group $SU(3, 2)$.

In chapter 7, we work with the two exceptional groups and prove Theorem F and Theorem 7.11. This depends upon a good amount of computer work in two respects. First, just describing the set $\mathcal{D}$ is a non-trivial exercise; see (2.12). Descriptions of $\mathcal{D}$ for the two exceptional groups are given in an Appendix (chapter 9). Second, describing the map $\mathbb{E}_\delta$ depends upon the orbit decompositions of [9]. We include the details for $G = E_{6,-14}$. These results finish the proof of Theorem A.

The discussion of various Loewy length estimates occurs in chapter 8.

Acknowledgements. The authors are grateful to Devra Garfinkle and Ken Johnson for discussions related to the combinatorial parametrizations of the sets $\mathcal{D}$. In addition, the second author would like to thank David Vogan for providing him, back in 1983, with his hand character calculations for the group $Sp_3\mathbb{R}$; in many ways, efforts to sort out those calculations provided a starting point for this manuscript.

2. NOTATION AND PRELIMINARIES

For the convenience of the reader, we quickly review some of the terminology and notation which will be in force throughout the paper. Once and for all, we fix a connected simple real matrix group of Hermitian symmetric type. For our purposes, there will be no harm in assuming G is among the groups in the following list, which includes "type labelings" used in the sequel:

(HS.1)	$SU(p,q), 1 \leq q \leq p$;
(HS.2)	$SO_e(2, 2n-1)$;
(HS.3)	$Sp_n\mathbb{R}$;
(HS.4)	$SO_e(2, 2n-2)$;
(HS.5)	$SO^*(2n)$;
(HS.6)	$E_{6,-14}$;
(HS.7)	$E_{7,-25}$.

Let θ be a Cartan involution for G, having fixed point set K. We are led to an Iwasawa decomposition $G{=}KAN$, a compatible minimal parabolic subgroup $P_m = M_m A_m N_m$ of G, real Lie algebras $\mathfrak{g}_o, \mathfrak{k}_o, \mathfrak{p}_{m,o}, \ldots$ and corresponding complexifications without subscript "o". If $P = MAN$ is the Langlands decomposition of a parabolic subgroup of G, then we reserve the notation M_{ss} to mean the semisimple factor of the reductive group M. Let $\mathbf{Q} = \mathbf{LU}$ be a parabolic subgroup of the complexification $\mathbf{G}$ of G; in general, we will reserve boldface capital roman letters for complex subgroups of a complexification $\mathbf{G}$ of G, whereas subgroups of G are denoted by plain capital roman letters. Fix a Borel subgroup $\mathbf{B}$ of $\mathbf{G}$; this determines systems of positive roots Φ^+ for the full root system and $\Phi^+_{\mathbf{Q}}$ for the roots of the Levi factor of $\mathbf{Q}$. Let W (resp. $W_{\mathbf{Q}}$) denote the Weyl group of $\mathfrak{g}$ (resp. $\mathfrak{l}$). We set

$$W^{\mathbf{Q}} = \{ w \in W \mid w^{-1}\Phi^+_{\mathbf{Q}} \subset \Phi^+ \},$$

which may be identified with the set of cosets $W_{\mathbf{Q}} \backslash W$, using the fact that each coset of $W_{\mathbf{Q}}$ in W will have a unique minimal length representative. With these identifications, the order relation $\leq$ on $W^{\mathbf{Q}}$ is the one inherited from the usual *Bruhat ordering* on W determined by $\mathbf{B}$, and the length function ℓ on W descends to one on $W^{\mathbf{Q}}$. The choice of a Borel subgroup $\mathbf{B}$ determines the set of simple reflections S generating W.

Both $\mathbf{K}$ and $\mathbf{Q}$ act with a finite number of orbits on the flag variety $\mathcal{B} = \mathbf{G}/\mathbf{B}$ of all Borel subalgebras of $\mathfrak{g}$. We will let V index the $\mathbf{K}$-orbits in $\mathcal{B}$ and recall that

$W^{\mathbf{Q}}$ parametrizes the $\mathbf{Q}$-orbits. In the sequel, we will need explicit parameters for the set V, but for now we simply note the $\mathbf{K}$- and $\mathbf{Q}$-orbit decompositions

$$\mathcal{B} = \bigcup_{w \in W^{\mathbf{Q}}} \mathcal{O}_w = \bigcup_{\delta_o \in V} \mathcal{V}_{\delta_o}. \tag{2.1}$$

We remark that under the one-to-one correspondence between $W^{\mathbf{Q}}$ and $\mathbf{Q}$-orbits, the Bruhat ordering corresponds to the relation given by orbit closure: i.e., $y \leq w$ if and only if $\mathcal{O}_y \subset \overline{\mathcal{O}_w}$. Likewise, the index set V carries an order relation $\preceq$, where $\gamma_o \preceq \delta_o$ if and only if $\mathcal{V}_{\gamma_o} \subset \overline{\mathcal{V}_{\delta_o}}$. This makes V into a partially ordered set; see [18] for elaboration.

Recall the category $\mathcal{HC}_o$ of Harish-Chandra modules with a fixed regular integral infinitesimal character has a finite number of distinct isomorphism classes of irreducible representations. Let $\mathcal{I}$ denote the finite set (up to equivalence) of irreducible Harish-Chandra modules in $\mathcal{HC}_o$. To each irreducible representation $\overline{\pi}$ in $\mathcal{I}$, we can associate its *Langlands data*: This consists of a triple $(P_\pi, \sigma_\pi, \nu_\pi)$, where $P_\pi = M_\pi A_\pi N_\pi$ is a cuspidal parabolic subgroup of G, σ_π a discrete series on M_π and ν_π a "positive character" on A_π. This data determines a generalized principal series $\mathrm{Ind}_{P_\pi}^G(\sigma_\pi \otimes \nu_\pi \otimes 1)$ having $\overline{\pi}$ as its unique irreducible quotient. Of course, keep in mind that we allow the cuspidal parabolic subgroup $P{=}G$; Langlands data involving $P = G$ corresponds to the discrete series representations of G (if such exist). As discussed in [23], $\mathcal{I}$ is in one-to-one correspondence with the set

$$\mathcal{D} = \{ (\mathcal{V}_{\delta_o}, \mathcal{L}_{\delta_o}) \mid \delta_o \in V \text{ and } \mathcal{L}_{\delta_o} \text{ is a } \mathbf{K}\text{-homogeneous line bundle on } \mathcal{V}_{\delta_o} \text{ with a flat connection} \}. \tag{2.2}$$

We will use the notation δ to denote a typical pair $(\mathcal{V}_{\delta_o}, \mathcal{L}_{\delta_o}) \in \mathcal{D}$. Of course, this is potentially confusing, since we also used greek letters for elements of the $\mathbf{K}$-orbit index set V; this explains the use of a subscript "o" (for "orbit") to distinguish between the two possible meanings. We point out in passing that $\mathcal{D}$ and V may or may not coincide; this will unfold more clearly in the sequel. We will use the notation $\overline{\pi}(\delta)$ to denote the irreducible Harish-Chandra module in $\mathcal{HC}_o$ corresonding to $\delta \in \mathcal{D}$. The representation $\overline{\pi}(\delta)$ occurs as the unique irreducible quotient of the generalized principal series representation

$$\pi(\delta) = \mathrm{Ind}_{P_{\pi(\delta)}}^G(\sigma_{\pi(\delta)} \otimes \nu_{\pi(\delta)} \otimes 1);$$

the representation $\pi(\delta)$ is referred to as a *standard module*. We now have two bases for the Grothendieck group of virtual characters of $\mathcal{HC}_o$:

$$\mathcal{I} = \{ \overline{\pi}(\delta) \mid \delta \in \mathcal{D} \} \quad \text{and} \quad \mathcal{S} = \{ \pi(\delta) \mid \delta \in \mathcal{D} \}. \tag{2.3}$$

For each standard Harish-Chandra module $\pi(\delta)$ we define

$$\ell^I(\pi(\delta)) = dim_{\mathbb{C}} \mathcal{V}_{\delta_o} - dim_{\mathbb{C}} \mathcal{B} + (1/2) dim_{\mathbb{R}} G/K.$$

This coincides with Vogan's definition [21, (8.1.4)]; one can check that the discrete series representations $\pi(\delta)$ have $\ell^I(\pi(\delta)) = 0$. Set $\ell^I(\delta) = \ell^I(\pi(\delta))$ for $\delta \in \mathcal{D}$.

Denote by $\mathrm{rad}^i(\pi(\delta))$ the ith-layer (i.e. the ith semisimple subquotient layer) of the radical filtration of $\pi(\delta)$, where $\mathrm{rad}^0(\pi(\delta)) = \overline{\pi}(\delta)$. A composition factor $\overline{\pi}(\gamma)$ of $\pi(\delta)$ is said to be *standard* if $\overline{\pi}(\gamma)$ lies in the $\ell^I(\pi(\delta)) - \ell^I(\pi(\gamma))$ layer of the radical filtration of $\pi(\delta)$.

(2.4) Pyramids: We now introduce a definition which is central to this paper. Recall the *G-order* $\prec_G$ of Lusztig-Vogan [16] on the set $\mathcal{D}$. If we are in a situation where $\mathcal{D} = V$ (or more generally $\mathcal{D}_o = V$, where $\mathcal{D}_o$ is the fundamental block as defined in (2.6) below), then we will work with the previously noted order relation $\preceq$ determined by **K**-orbit closure. Given a maximal cuspidal parabolic subgroup $P \subset G$ we define subsets of $\mathcal{D}$ as follows:

$$\text{(2.4a)} \quad \begin{aligned} \mathcal{D}^{\max} &= \{\delta \in \mathcal{D} \mid \text{the Langlands data for } \pi(\delta) \text{ involves a} \\ &\qquad\qquad \text{maximal parabolic subgroup}\} \\ \mathcal{D}^{\max}_P &= \{\delta \in \mathcal{D}^{\max} \mid \text{the Langlands data for } \pi(\delta) \text{ involves } P\}. \end{aligned}$$

This leads to the decomposition

$$\text{(2.4b)} \qquad \mathcal{D}^{\max} = \bigcup\nolimits_P \mathcal{D}^{\max}_P.$$

Given a *maximal* cuspidal parabolic subgroup $P = MAN$, denote by M_{ds} the set of equivalence classes of discrete series of M having a *fixed* central and regular integral infinitesimal character; if M is connected, then this set is parametrized by a coset space $W_{\mathbf{P}}^{\mathbf{K_M}} = W_{\mathbf{K_M}} \backslash W_{\mathbf{P}}$, where $\mathbf{K_M} = \mathbf{K} \cap \mathbf{M}$ is a complexified maximal compact subgroup of M. Given $\sigma \in M_{\mathrm{ds}}$ and τ a discrete series representation of M, we say τ is of *type* σ, if τ corresponds to σ under a translation functor equivalence of categories for the category of Harish-Chandra modules on M. Define

$$\text{(2.4c)} \qquad \mathcal{P}_{\sigma,M} = \{\delta \in \mathcal{D}^{\max}_P \mid \text{the inducing data for } \pi(\delta) \text{ is of type } \sigma \text{ on } M\}.$$

We refer to $\mathcal{P}_{\sigma,M}$ as a *pyramid* and we have the decomposition

$$\text{(2.4d)} \qquad \mathcal{D}^{\max}_P = \bigcup_{\sigma \in M_{\mathrm{ds}}} \mathcal{P}_{\sigma,M}.$$

The Bruhat poset $W^{\mathbf{P}}$ carries a *Poincaré involution* ξ which leads to

$$\text{(2.4e)} \qquad W^{\mathbf{P}}_{\text{upper}} = \{w \in W^{\mathbf{P}} \mid \ell(\xi(w)) \leq \ell(w)\},$$

which is just the "upper half" of $W^{\mathbf{P}}$. Define a shifted length function ℓ' on $W^{\mathbf{P}}_{\text{upper}}$ by the rule $\ell' = \ell - c$, where c is chosen so that the minimal elements of (2.4e) have length 1. Recall

LEMMA 2.4f [5, §3]. *The set $\mathcal{P}_{\sigma,M}$ together with the inherited weak $\preceq_w$ order and length function ℓ^I is poset isomorphic with $W^{\mathbf{P}}_{\text{upper}}$ together with its weak order and length function ℓ'.*

These posets have been studied in great detail, since the categories $\mathcal{O}(\mathfrak{g},\mathfrak{p})$ (for $\mathfrak{p}$ the complexified Lie algebra of a maximal cuspidal parabolic subgroup $P \subset G$) are of multiplicity free type, as studied in [2] and [7]. Each pyramid $\mathcal{P}_{\sigma,M}$ will possess a unique element of maximal length, denoted $\delta_{\sigma,M}$, allowing us to define

(2.4g) $$\widehat{\mathcal{P}}_{\sigma,M} = \{\, \delta \in \mathcal{D} \,|\, \delta \preceq_G \delta_{\sigma,M} \,\}$$

referred to as a *G-pyramid*. As a preliminary orientation for the reader, we offer two examples of the above ideas; several others will come in the sequel:

(2.4h) For a real rank one group, [7, Figs. 4.3–4.6] describes the G-pyramids. If $G \neq SL_2\mathbb{R}$, then M_m is compact and connected and there is exactly one pyramid, obtained by removing the bottom row of the cited diagrams; compare to [7, Figs. 8.4–8.7] for the $W^{\mathbf{P}_m}$ and $W^{\mathbf{P}_m}_{\text{upper}}$ diagrams. In this case, the minimal parabolic subgroup is the unique (up to conjugacy) maximal cuspidal parabolic subgroup of G, so every pyramid is of type $\mathcal{P}_{\sigma,M_m}$. In the case of $SL_2\mathbb{R}$, M_m is compact with two components, so there are two pyramids, each consisting of a single parameter; one indexing the principal series with finite dimensional quotient in $\mathcal{HC}_o$ and the other corresponding to the irreducible principal series in $\mathcal{HC}_o$; see (2.8) below.

(2.4i) For the real rank two group *SU(2,2)*, there is a unique (up to conjugacy) maximal cuspidal parabolic subgroup $P = MAN$, having $M_{ss} = SL_2\mathbb{R}$. Consequently, there will be two pyramids, denoted $\mathcal{P}_{\pm,M}$. Refer to [9, Table 2] for a description of the set V. As remarked in (2.7) below, V parametrizes the fundamental block in $\mathcal{D}$. The set of Langlands data attached to P will correspond to the following subset of V:

$$\{\delta_i \mid i \notin \{21, 19, 15, 6, 5, 4, 3, 2, 1\}\}$$

The form of $W^{\mathbf{P}}_{\text{upper}}$ is determined by [7, Fig. 8.5], with $n = 3$. One finds that

$$\mathcal{P}_{+,M} = \{\delta_i \,|\, i = 20, 17, 16, 12, 11, 9\},$$
$$\mathcal{P}_{-,M} = \{\delta_i \,|\, i = 18, 14, 13, 10, 8, 7\},$$
$$\hat{\mathcal{P}}_{+,M} = \mathcal{P}_{+,M} \cup \{\delta_i \,|\, i = 15, 10, 8, 6, 5, 4, 3, 2\,\},$$

and

$$\hat{\mathcal{P}}_{-,M} = \mathcal{P}_{-,M} \cup \{\delta_i \,|\, i = 15, 12, 9, 6, 5, 3, 2, 1\,\}.$$

(2.5) Hecke modules: We will need to assume the reader is familiar with the theory of [5], [23] and [24]. Let's briefly recall some of the basic ideas. First, each of the categories $\mathcal{HC}_o$ and $\mathcal{O}(\mathfrak{g}, \mathfrak{q}^\sharp)$ gives rise to a module over the *Hecke algebra* $\mathcal{H}$ associated to the Coxeter pair (W, S). In the Harish-Chandra module setting, we define a free $\mathbb{Z}[u^{1/2}, u^{-1/2}]$-module

$$\mathcal{M}_{\mathbf{K}} = \mathbb{Z}[u^{1/2}, u^{-1/2}] \otimes_{\mathbb{Z}} \mathbb{Z}[\mathcal{D}] \tag{2.5a}$$

as in [23, (6.4)]. Likewise, for the parabolic subgroup $\mathbf{Q}^\sharp$ of $\mathbf{G}$ (defined in the introduction) we construct the module

$$\mathcal{M}_{\mathbf{Q}^\sharp} = \mathbb{Z}[u^{1/2}, u^{-1/2}] \otimes_{\mathbb{Z}} \mathbb{Z}[W^{\mathbf{Q}^\sharp}] \tag{2.5b}$$

as in [4, (3.5)]. Each of $\mathcal{M}_{\mathbf{K}}$ and $\mathcal{M}_{\mathbf{Q}^\sharp}$ becomes a module over $\mathcal{H}$, as described in the cited references, and the combinatorial baggage associated to these modules (again described in these references) controls the character theory of the categories $\mathcal{HC}_o$ and $\mathcal{O}(\mathfrak{g}, \mathfrak{q}^\sharp)$, respectively.

We especially want to recall the notion of the Hecke algebra operator T_s being of *compact type, complex type, real type not satisfying the parity condition, real type I, real type II, noncompact type I or noncompact type II*. This corresponds to the type of simple root through which the abstract reflection $s \in S$ is defined. These definitions are reviewed in detail for $G = SU(p,q)$ in (2.7). We have the refined notions of *upward* and *downward complex type*, corresponding to cases [23, (6.4b1)] and [23, (6.4b2)], respectively, in the Harish-Chandra setting and cases [4, (3.12b)] and [4, (3.12c)], respectively, in the $\mathcal{O}(\mathfrak{g}, \mathfrak{q}^\sharp)$ setting. Also, we recall the *cross* and *circle actions* of $\mathcal{H}$, denoted $\times$ and $\circ$, on the module $\mathcal{M}_{\mathbf{K}}$, as defined in [23, §6]. For example, if $\delta \in \mathcal{D}$ parametrizes a discrete series representation and $s \in S$ is of noncompact type I for δ, then the Cayley transform of δ through s, denoted $s \circ \delta$, parametrizes a module in $\mathcal{D}^{\max}$ and $\pi(s \circ \delta)$ (a *character identity*) has radical filtration involving three irreducible composition factors: the discrete series modules $\overline{\pi}(\delta)$ and $\overline{\pi}(s \times \delta)$ together with $\overline{\pi}(s \circ \delta)$. The radical filtration of $\pi(s \circ \delta)$ has the structure:

$\overline{\pi}(s \circ \delta)$
$\overline{\pi}(\delta) \oplus \overline{\pi}(s \times \delta)$

We further recall the following:

(2.5c) We have sets of polynomials $\{Q_{\gamma,\delta}(u) \in \mathbb{Z}[u] \mid \gamma, \delta \in \mathcal{D}\}$ and $\{P_{y,w}(u) \in \mathbb{Z}[u] \mid y, w \in W^{\mathbf{Q}^\sharp}\}$ associated to these two Hecke modules via [23, (6.12)] and [4, (3.21)]; i.e., the Lusztig-Vogan and relative Kazhdan-Lusztig polynomials.

(2.5d) If $\mathbb{D}$ denotes the natural duality on our Hecke modules, then the sets $\{\widehat{C_\delta} \mid \delta \in \mathcal{D}\}$ and $\{\widehat{C_w} \mid w \in W^{\mathbf{Q}^\sharp}\}$ form $\mathbb{D}$ self-dual bases for the Hecke modules $\mathcal{M}_{\mathbf{K}}$ and $\mathcal{M}_{\mathbf{Q}^\sharp}$, respectively; this is explained in [23, (6.12)] and [4, (3.21)].

The connection between the combinatorics of (2.5c, d) and character theory proceeds via geometry, as described in [4], [5] or [16]. We will assume full familiarity with these concepts, but emphasize one key circle of ideas. The Hecke module formulas

$$\widehat{C_\delta} = u^{-\ell(\delta)/2} \sum_{\gamma \preceq_G \delta} Q_{\gamma,\delta}(u)\gamma \tag{2.5e}$$

and

$$\widehat{C_w} = u^{-\ell(w)/2} \sum_{y \leq w} P_{y,w}(u)y \tag{2.5f}$$

in $\mathcal{M}_{\mathbf{K}}$ and $\mathcal{M}_{\mathbf{Q}^\sharp}$ correspond to Grothendieck group formulas (for sheaf categories $\mathcal{E}(\mathbf{K})'$ and $\mathcal{E}(\mathbf{Q}^\sharp)'$ defined as in [5, §5])

$$[IC(\mathcal{V}_\delta)] = L^{-\ell(\delta)/2} \otimes \sum_{\gamma \preceq_G \delta} Q_{\gamma,\delta}(u)\mathbb{I}_{\mathcal{V}_\gamma} \tag{2.5g}$$

and

$$[IC(\mathcal{O}_w)] = L^{-\ell(w)/2} \otimes \sum_{y \leq w} P_{y,w}(u)\mathbb{I}_{\mathcal{O}_y} \tag{2.5h}$$

expressing the Euler characteristic (denoted [...]) of the *intersection cohomology complexes* $IC(\mathcal{V}_\delta)$ and $IC(\mathcal{O}_w)$ attached to the $\mathbf{K}$-orbit closure $\bar{\mathcal{V}}_\delta$ and the $\mathbf{Q}^\sharp$-orbit closure $\bar{\mathcal{O}}_w$ as a sum of *Tate twisted* (denoted $L^{i/2}$) locally constant $\mathbf{K}$- or $\mathbf{Q}^\sharp$-equivariant sheaves $\mathbb{I}_{\mathcal{V}_\delta}$ or $\mathbb{I}_{\mathcal{O}_w}$ on the various orbits. (Note: There is some possible confusion here, since several different δ's may be associated to the same $\mathbf{K}$-orbit. If δ and δ' are associated to the same orbit $\mathcal{V}_{\delta_o}$ then two different local systems are being indexed and hence two different $\mathbf{K}$-equivariant locally constant sheaves on the same orbit.) Inverting the unipotent upper triangular matrices $(Q_{\gamma,\delta}(u))$ and $(P_{y,w}(u))$ will "invert" the formulas in (2.5e-h), leading to Hecke module formulas

$$\delta = \sum_{\gamma \preceq_G \delta} S_{\gamma,\delta}(u)\widehat{C_\gamma} \tag{2.5i}$$

and

$$w = \sum_{y \leq w} S^\natural_{y,w}(u)\widehat{C_y} \tag{2.5j}$$

which correspond to the sheaf theoretic formulas

$$\mathbb{I}_{\mathcal{V}_\delta} = \sum_{\gamma \preceq_G \delta} S_{\gamma,\delta}(u)[IC(\mathcal{V}_\gamma)] \tag{2.5k}$$

and

$$\mathbb{I}_{\mathcal{O}_w} = \sum_{y \leq w} S^{\natural}_{y,w}(u)[IC(\mathcal{O}_y)] \tag{2.5l}$$

Computing the Laurent polynomials $S_{\gamma,\delta}, S^{\natural}_{y,w} \in \mathbb{Z}[u^{1/2}, u^{-1/2}]$ is tantamount to obtaining closed formulas for the radical filtrations of the standard modules $\pi(\delta)$ in $\mathcal{HC}_o$ and the generalized Verma modules in $\mathcal{O}(\mathfrak{g}, \mathfrak{q}^{\sharp})$. This is because a term of the form $u^{i/2}\widehat{C_\gamma}$ in (2.5i) will imply the existence of $\overline{\pi}(\gamma)$ in the ith-layer of the weight filtration of $\pi(\delta)$ and (up to a shift in indexing), the radical and weight filtrations coincide by a result of Casian [3]; similar remarks hold in the category $\mathcal{O}(\mathfrak{g}, \mathfrak{q}^{\sharp})$ setting.

(2.6) The Stucture of $\mathcal{D}$: The parameter set $\mathcal{D}$ can be decomposed as a union $\mathcal{D} = \cup \mathcal{D}_i$, where $\mathcal{D}_i$ denote the *blocks* of $\mathcal{D}$; see [21, §9.2]. Fix a finite dimensional representation F of G and recall we are assuming G is connected. We will use the notation $\mathcal{D}_o$ to denote the *fundamental block*; i.e. the block containing the parameter δ_F attached to F. This allows us to define

$$\mathcal{D}_o^{\max} = \mathcal{D}^{\max} \cap \mathcal{D}_o. \tag{2.6a}$$

The block decomposition of $\mathcal{D}$ will determine an $\mathcal{H}$-invariant decomposition

$$\mathcal{M}_{\mathbf{K}} = \bigoplus \mathcal{M}_{\mathbf{K},i},$$

where each $\mathcal{M}_{\mathbf{K},i}$ is a Hecke submodule of $\mathcal{M}_{\mathbf{K}}$.

In sections (2.7)–(2.12), we will parametrize the fundamental block $\mathcal{D}_o$. The procedure for obtainining these results is explained in detail in [24]; we assume familiarity with this theory. Let's emphasize a couple of basic points. First, let $\{H_1, \ldots, H_k\}$ be a collection of θ-stable representatives for the conjugacy classes of Cartan subgroups in G. Denote by $H_{i,o}$ the identity component of H_i and define

$$W(G, H_i) = (\text{Normalizer in}\, G\, \text{of}\, H_i)/H_i \tag{2.6b}$$

to be the Weyl group of H_i in G. One can count that

$$|\mathcal{D}| = \sum_{1 \leq i \leq k} |H_i/H_{i,o}|\, |W/W(G, H_i)| \tag{2.6c}$$

Consequently, the starting point in parametrizing $\mathcal{D}$ or $\mathcal{D}_o$ is the determination of Cartan subgroup conjugacy classes, the component groups of the Cartan subgroups and their Weyl groups in G. The conjugacy classes are described in [19] and the other information depends on [24, (4.16)]. With this data in hand, the theory in [24, §§1–8] leads to the description of the $\preceq_G$ order on $\mathcal{D}$. In what follows, we will either simply state a parametrization and give a reference, or alternatively provide the crucial data, leaving verification to the reader. In the exceptional cases, this becomes a highly non-trivial exercise, even with the aid of a computer.

Finally, given any θ-stable Cartan subgroup H, we can write $H = TA$, where $T = K \cap H$ is compact and A is a vector group; the decomposition arises through the ± 1-eigenspaces of θ. Given $H = TA$, we let L denote the centralizer of A in G; then $L = MA$ becomes the Levi factor of a cuspidal parabolic subgroup $P = MAN$, where T is a compact Cartan subgroup of M.

(2.7) Explicit parameter set for type HS.1: Throughout this subsection, assume that $G = SU(p,q)$ with $p \geq q \geq 1$, and set $n = p + q$. Recall the Satake diagram has the form

(2.7a)

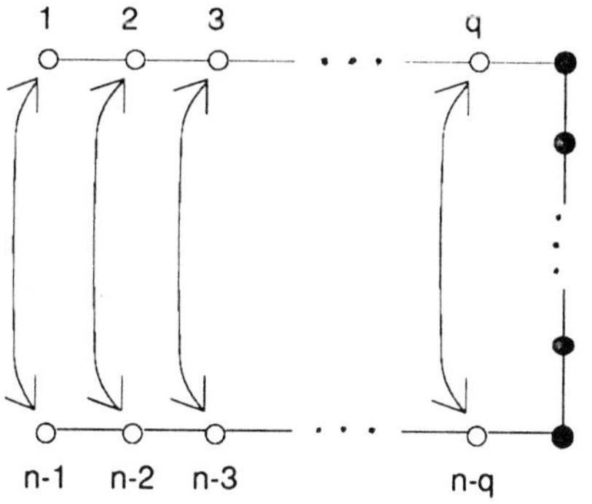

A combinatorial parametrization of the set $\mathcal{D}_o$ is described in [12] as follows. (Essentially the same parametrization appears in a different guise in [17].) To each element of $\mathcal{D}_o$ is associated an arrangement δ of the numbers $1, 2, \cdots, n$ in singles or pairs. Each single number has a $+$ sign or a $-$ sign attached to it. Let r be the number of pairs in δ; then we require that $0 \leq r \leq q$, and that the number of $+$ signs in δ equal $p - r$. Although the ordering of the single numbers, pairs, and of elements within the pairs in δ is immaterial, we shall always write parameters in the following "standard ordering": the sequence of numbers obtained by deleting the second coordinate of each pair in δ, should increase from left to right; and the second coordinate of each pair in δ should be greater than the corresponding first coordinate. For example, a typical δ for $SU(4,3)$ is $\delta = 1^+(25)(34)6^-7^+$; other examples occur in Table 6.1.

Next we recall the action of simple root reflections on $\mathcal{D}_o$. Put $\alpha_i = e_i - e_{i+1}$, $1 \leq i \leq n-1$, the usual simple roots, and let Φ^+ be the corresponding set of positive roots. To each $\delta \in \mathcal{D}_o$ (parametrized as above) we associate a Cartan involution θ_δ defined on $\mathfrak{h}^* = \{\sum_{i=1}^{n} c_i e_i | c_i \in \mathbb{C}, \sum c_i = 0\}$ as follows:

$$\theta_\delta(e_i) = \begin{cases} e_i & \text{if } \; i^\epsilon \in \delta, \text{ for some } \epsilon \in \{+,-\} \\ e_j & \text{if } \; (i,j) \in \delta\,. \end{cases} \tag{2.7b}$$

We say that a simple root α is *real*, *imaginary*, or *complex* for δ if $\theta_\delta(\alpha) = -\alpha$, $\theta_\delta(\alpha) = \alpha$, or $\theta_\delta(\alpha) \notin \{\alpha, -\alpha\}$, respectively. If α_i is imaginary for δ, we specify futher that α_i is

(2.7c1) *compact imaginary* for δ if $i^\epsilon, (i+1)^\epsilon \in \delta$ for some $\epsilon \in \{+,-\}$

(c2) *noncompact imaginary* for δ if $i^\epsilon, (i+1)^{-\epsilon} \in \delta$ for some $\epsilon \in \{+,-\}$.

If α is complex for δ, we say that α is

(c3) complex of *upward type* for δ if $\theta_\delta(\alpha) \in \Phi^+$
(c4) complex of *downward type* for δ if $\theta_\delta(\alpha) \in -\Phi^+$.

Let s_i be the reflection corresponding to the simple root α_i, $1 \leq i \leq n-1$. For $\delta \in \mathcal{D}_o$,

(2.7d1) if α_i is complex for δ, $s_i \circ \delta = s_i \times \delta$ is obtained by interchanging i and $i+1$ in δ;
(d2) if α_i is noncompact imaginary for δ, $s_i \circ \delta$ is obtained by replacing the two signed entries $i^\epsilon, (i+1)^{-\epsilon}$ by the pair $(i, i+1)$;
(d3) if α_i is compact imaginary for δ, $s_i \times \delta = \delta$;
(d4) if α_i is real for δ, $s_i \circ \delta$ consists of the *two* parameters obtained by replacing the pair $(i, i+1)$ by the two signed entries $i^+(i+1)^-$ or by the two signed entries $i^-(i+1)^+$.

If α_i is complex of upward type or noncompact imaginary for δ, write $s_i \circ \delta \xrightarrow{s_i} \delta$; then $\delta \preceq_w s_i \circ \delta$ in the *weak order* on $\mathcal{D}_o$, where the weak order $\preceq_w$ is by definition the smallest partial order containing these relations.

Finally recall that the *Bruhat G-order* is the smallest partial order $\preceq_G$ on $\mathcal{D}_o$ including the weak order and having the property: for $\gamma, \delta \in \mathcal{D}_o$, if there exists a simple reflection $s \in S$ and $\gamma', \delta' \in \mathcal{D}_o$ with $\gamma \xrightarrow{s} \gamma', \delta \xrightarrow{s} \delta'$, and $\gamma' \preceq_G \delta'$, then $\gamma \preceq_G \delta$. We remark that in the case $G = SU(p,q)$ under consideration, the set $\mathcal{D}_o$ coincides with the set V of (2.1), and the two orderings $\preceq_G$ and $\preceq$ also coincide. Henceforth we will use the simpler notation $\preceq$ for the G-order. (Caution: For $SU(p,p)$, $\mathcal{D} \neq \mathcal{D}_o$ and one must proceed more carefully; this will be clarified in the sequel.)

(2.8) Explicit parameter set for type HS.2: In this case, $G = SO_e(2, 2n-1)$, the identity component of $SO(2, 2n-1)$. (We use the subscript "e" rather than "0" to conform with the classical notation of [14].) Recall the Satake diagram

(2.8a)
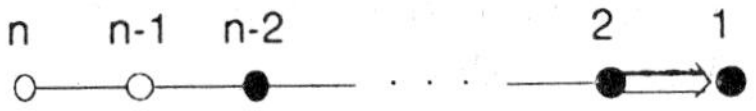

By [19], there are four conjugacy classes of Cartan subgroups in G. Denote their associated cuspidal parabolic subgroups $P = MAN$ (as in (2.6)) of G; note that the cuspidal parabolic subgroup associated to a compact Cartan subgroup is just G:

(2.8b)
$$\begin{aligned} H_m &= T_m A_m = \text{a maximally split Cartan subgroup;} \\ H_n &= T_n A_n = \text{a Cartan subgroup with } \quad \alpha_n \in \Phi_{P_n}; \\ H_{n-1} &= T_{n-1} A_{n-1} = \text{a Cartan subgroup with } \quad \alpha_{n-1} \in \Phi_{P_{n-1}}; \\ H_c &= T_c A_c = \text{a compact Cartan subgroup.} \end{aligned}$$

Define

(2.8c)
$$\mathcal{D}_{P_m}, \mathcal{D}_{P_n}, \mathcal{D}_{P_{n-1}} \text{ and } \mathcal{D}_G$$

as in (2.4a). Then

(2.8d) $$\mathcal{D} = \mathcal{D}_{P_m} \cup \mathcal{D}_{P_n} \cup \mathcal{D}_{P_{n-1}} \cup \mathcal{D}_G.$$

We now use the theory of [24] to arrive at the data in Table 2.1; in particular, this determines the cardinality of $\mathcal{D}$ to be $3n^2$. As a special case, we remark that $SO_e(2,3)$ is isomorphic to the adjoint group of $Sp_2\mathbb{R}$ and the "$\mathcal{D}$ set" for this latter group is given in [22].

TABLE 2.1. Cartan data for $SO_e(2, 2n-1)$

H	$\dim A$	P	M_{ss}	$\lvert H/H_o\rvert$	$\lvert W(G,H)\rvert$	$\lvert\mathcal{D}_P\rvert$
H_m	2	P_m	$SO(2n-3)$	2	$2^{n+1}(n-2)!$	$n(n-1)$
H_n	1	P_n	$SL_2\mathbb{R} \times SO(2n-3)$	1	$2^{n-1}(n-2)!$	$2n(n-1)$
H_{n-1}	1	P_{n-1}	$SO(1,2n-2)$	1	$2^n(n-1)!$	n
H_c	0	G	G	1	$2^{n-1}(n-1)!$	$2n$

It remains to describe the $\preceq_w$ order on $\mathcal{D}$. To this end, we first parametrize the sets in (2.8c). We define

(2.8e) $$\mathcal{T}^\sharp = \{(i,j)^\sharp \mid 1 \le i \le n-1, 1 \le j \le n-i\}$$

and make this into a labeled graph using only the labeled edges in Figure 2.2 that connect parameters within the set $\mathcal{T}^\sharp$. Define another labeled graph, denoted $\mathcal{T}^\flat$, by replacing all occurrences of $\sharp$ by $\flat$. Using [24] and Table 2.1, we find that

(2.8f) $$\mathcal{D}_{P_m} = \mathcal{T}^\sharp \cup \mathcal{T}^\flat,$$

corresponding to the fact $|(M_m)_{\mathrm{ds}}| = 2$. With the ordering imposed above, this gives us all of the complex type weak order relations on $\mathcal{D}_{P_m}$. In addition, the parameter set $\mathcal{T}^\flat$ admits real type reflections not satisfying the parity condition, denoted by loops in Figure 2.2. Next, using Table 2.1 and (2.6), $|(M_n)_{\mathrm{ds}}| = 2$. Consequently,

(2.8g) $$\begin{aligned} \mathcal{D}_{P_n} &= \mathcal{P}_{+,M_n} \cup \mathcal{P}_{-,M_n}; \\ \mathcal{P}_{\pm,M_n} &= W^{\mathbf{P}_n}_{\mathrm{upper}}. \end{aligned}$$

The pyramids in (2.8g) are the upper halves of $(B_n, A_1 \times B_{n-2})$ Hasse diagrams. Such diagrams are parametrized as the upper half of [7, Fig. 8.6]; one must relabel the Dynkin diagram of this reference to be compatible with (2.8a) and also replace "$n+1$" by "n". Thus, we may parametrize the pyramids by pairs

(2.8h) $$\begin{aligned} \mathcal{P}_{+,M_n} &= \{(i,j) \mid 0 \le i \le n-2,\ i+1 \le j \le 2n-2-i\}; \\ \mathcal{P}_{-,M_n} &= \{\overline{(i,j)} \mid 0 \le i \le n-2,\ i+1 \le j \le 2n-2-i\}, \end{aligned}$$

together with the complex type weak $\preceq_w$ order relations given by the cited reference. Again, using Table 2.1 and (2.6), $|(M_{n-1})_{\text{ds}}| = 1$. Consequently,

$$\text{(2.8i)} \qquad \begin{aligned} \mathcal{D}_{P_{n-1}} &= \mathcal{P}_{+,M_{n-1}}; \\ \mathcal{P}_{+,M_{n-1}} &= W^{\mathbf{P}_{n-1}}_{\text{upper}}. \end{aligned}$$

The pyramids in (2.8i) are the upper halves of (B_n, B_{n-1}) Hasse diagrams. Such diagrams are parametrized as

$$\text{(2.8j)} \qquad \mathcal{D}_{P_{n-1}} = \mathcal{P}_{+,M_{n-1}} = \{\hat{i} \mid 1 \le i \le n\},$$

together with the complex type weak order relations given by: $s_{i+1} \times \delta_{\hat{i}} = \delta_{\widehat{i+1}}$, $1 \le i \le n-1$. Another application of Table 2.1 and (2.6) shows that the discrete series representations for G are parametrized by

$$\text{(2.8k)} \qquad \mathcal{D}_G = \{i^* \mid 1 \le i \le 2n\}.$$

In Figures 2.1 and 2.2, we indicate the weak $\preceq_w$ order relations among the four subsets of (2.8c). In Figure 2.1, the order relations emanating from the bottom row are noncompact Type I and the edges labeled "1" emanating from $\hat{i}$, $2 \le i \le n$ are noncompact Type II attaching to the indicated elements of $\mathcal{D}_{P_m}$ In Figure 2.2, we show these same noncompact Type II relations, the complex type weak order relations of $\mathcal{D}_{P_m}$, and the real Type I relations connecting $\mathcal{D}_{P_m}$ to $\mathcal{D}_{P_n}$. The real type reflections failing the *parity condition* are denoted by labeled loops in the figure. Finally, after pasting these two pictures together, this gives all the weak order relations. In particular, if a number "i" does not label an edge or loop emanating from a node δ in $\mathcal{D}$, then s_i is of compact type for the node δ.

We have shown that $\mathcal{D}$ coincides with the fundamental block. In Figure 2.3, we illustrate the situation with the diagram associated to $SO_e(2,7)$. As a second example, we have already noted that $PSp_2\mathbb{R} = SO_e(2,3)$ and the corresponding diagram appears (inverted) in [22].

(2.9) Explicit parameter set for type HS.3: In this case, $G = Sp_n\mathbb{R}$. We label the Satake diagram as follows

(2.9a)

This group admits a number of blocks, one for each pair of integers (r, s) satisfying: $r \ge s$ and $r + s = 2n + 1$. To the extent necessary, the blocks are described in chapter 3.

(2.10) Explicit parameter set for type HS.4: In this case, $G = SO_e(2, 2n-2)$ and the Satake diagram is

(2.10a)

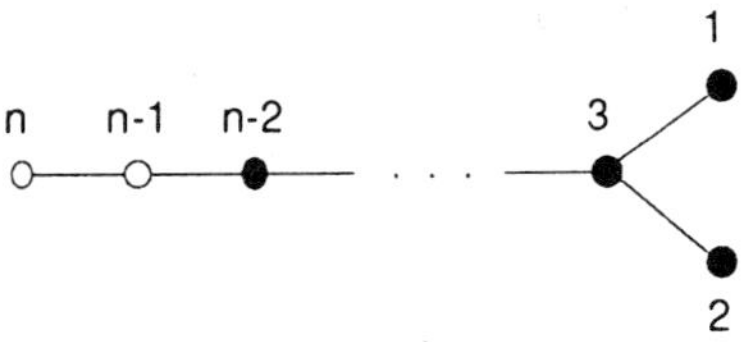

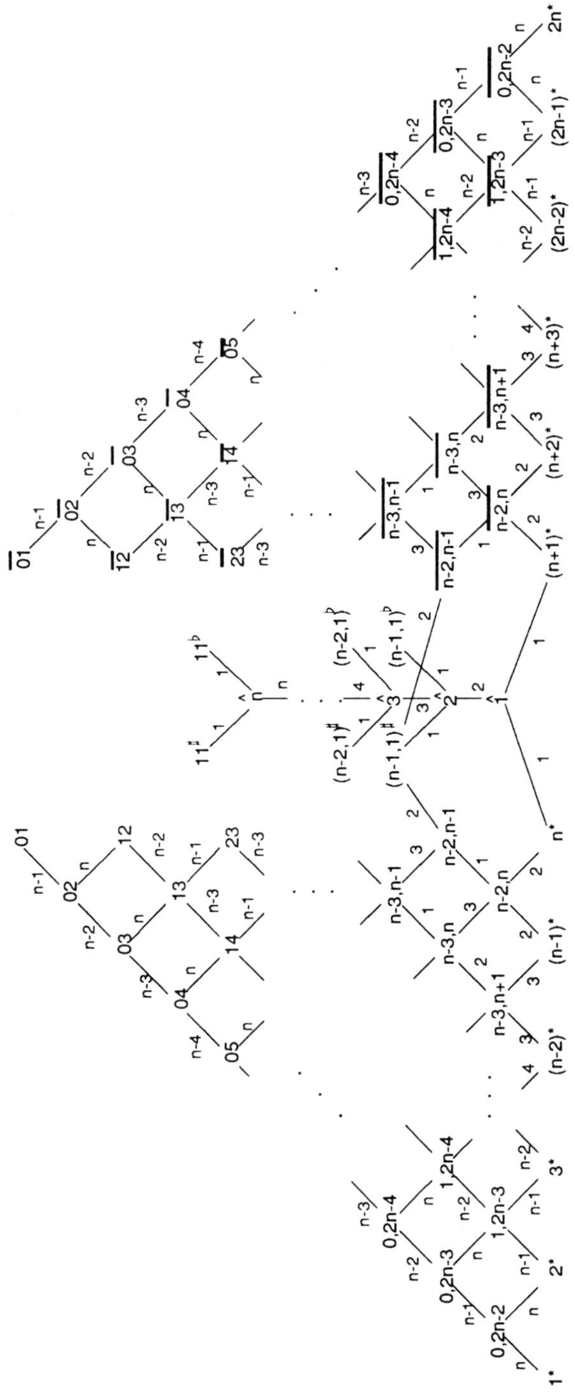

FIGURE 2.1. Part I of $\mathcal{D}$ for $SO_e(2, 2n-1)$

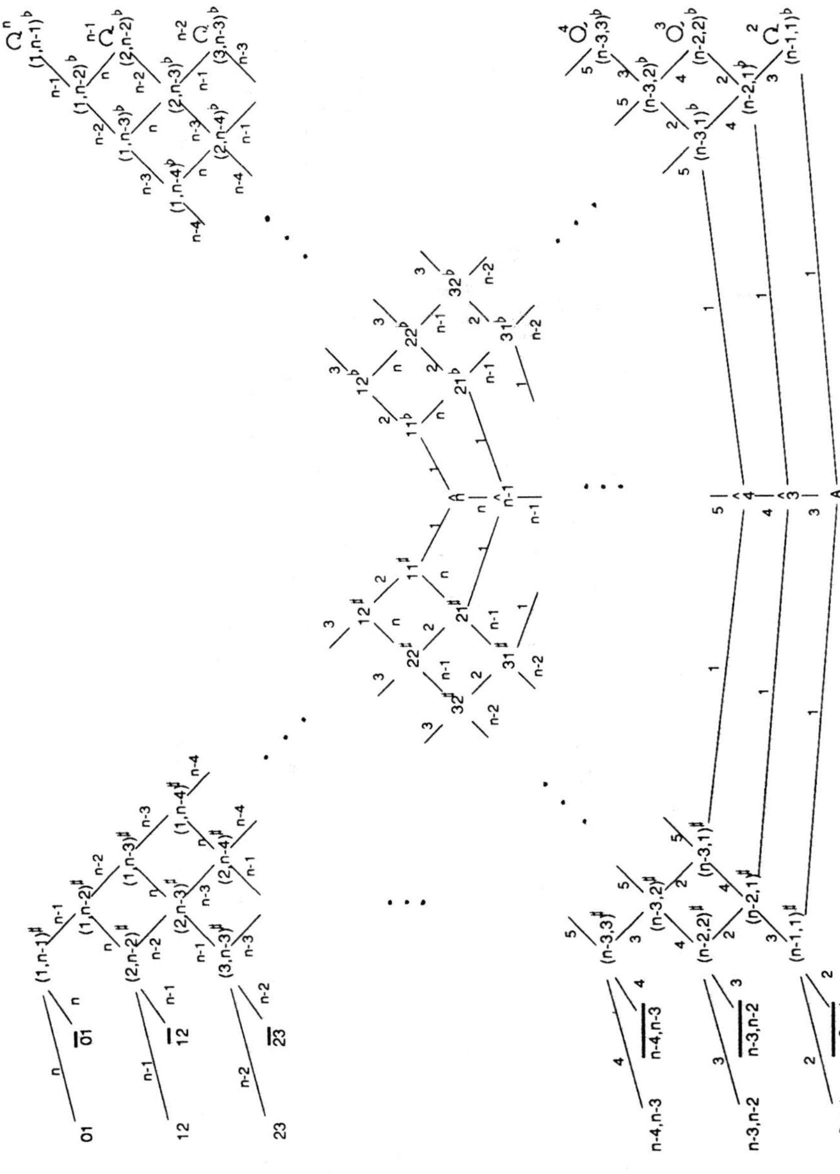

FIGURE 2.2. Part II of $\mathcal{D}$ for $SO_e(2, 2n-1)$

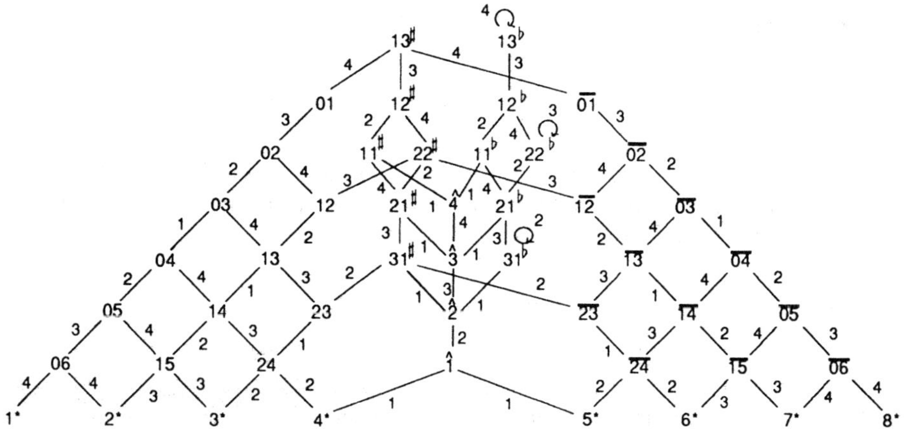

FIGURE 2.3. $\mathcal{D}$ for $SO_e(2,7)$

By [19], there are three conjugacy classes of Cartan subgroups in G and their associated cuspidal parabolic subgroups $P = MAN$ (as in (2.6)) of G; note that the cuspidal parabolic subgroup associated to a compact Cartan subgroup is just G:

$$
\begin{aligned}
&H_m = T_m A_m = \text{a maximally split Cartan subgroup;} \\
(2.10\text{b}) \qquad &H_n = T_n A_n = \text{a Cartan subgroup with} \quad \alpha_n \in \Phi_{P_n}; \\
&H_c = T_c A_c = \text{a compact Cartan subgroup.}
\end{aligned}
$$

Define

$$
(2.10\text{c}) \qquad \mathcal{D}_{P_m}, \mathcal{D}_{P_n} \text{ and } \mathcal{D}_G
$$

as in (2.4a). Then

$$
(2.10\text{d}) \qquad \mathcal{D} = \mathcal{D}_{P_m} \cup \mathcal{D}_{P_n} \cup \mathcal{D}_G.
$$

We now use the theory of [24] to arrive at the data in Table 2.2; in particular, this determines the cardinality of $\mathcal{D}$ to be $3n(n-1)$.

TABLE 2.2. Cartan data for $SO_e(2, 2n-2)$

H	$\dim A$	P	M_{ss}	$\lvert H/H_o\rvert$	$\lvert W(G,H)\rvert$	$\lvert\mathcal{D}_P\rvert$
H_m	2	P_m	$SO(2n-4)$	2	$2^n(n-2)!$	$n(n-1)$
H_n	1	P_n	$SL_2\mathbb{R} \times SO(2n-4)$	1	$2^{n-2}(n-2)!$	$2n(n-1)$
H_c	0	G	G	1	$2^{n-2}(n-1)!$	$2n$

We first parmetrize the sets in (2.10c). We define

$$\mathcal{T}^\sharp = \{(i,j)^\sharp \mid 1 \le i \le n-1,\ 1 \le j \le n-i\} \tag{2.10e}$$

and make this into a labeled graph using the labeled edges in Figure 2.4. Define another labeled graph, denoted $\mathcal{T}^\flat$, by replacing all occurrences of $\sharp$ by $\flat$. Using [24] and Table 2.2, we find that

$$\mathcal{D}_{P_m} = \mathcal{T}^\sharp \cup \mathcal{T}^\flat; \tag{2.10f}$$

the ordering imposed above gives us all of the complex type weak order relations on $\mathcal{D}_{P_m}$. In this case, the parameter set $\mathcal{T}^\flat$ is itself a block, denoted $\mathcal{D}_1$, admitting real type reflections not satisfying the parity condition, denoted by loops in Figure 2.4. With the convention that unaccounted for edge labels are of compact type, Figure 2.4 describes the ordered set $\mathcal{D}_1$ (together with part of $\mathcal{D}_o$, as will be explained below).

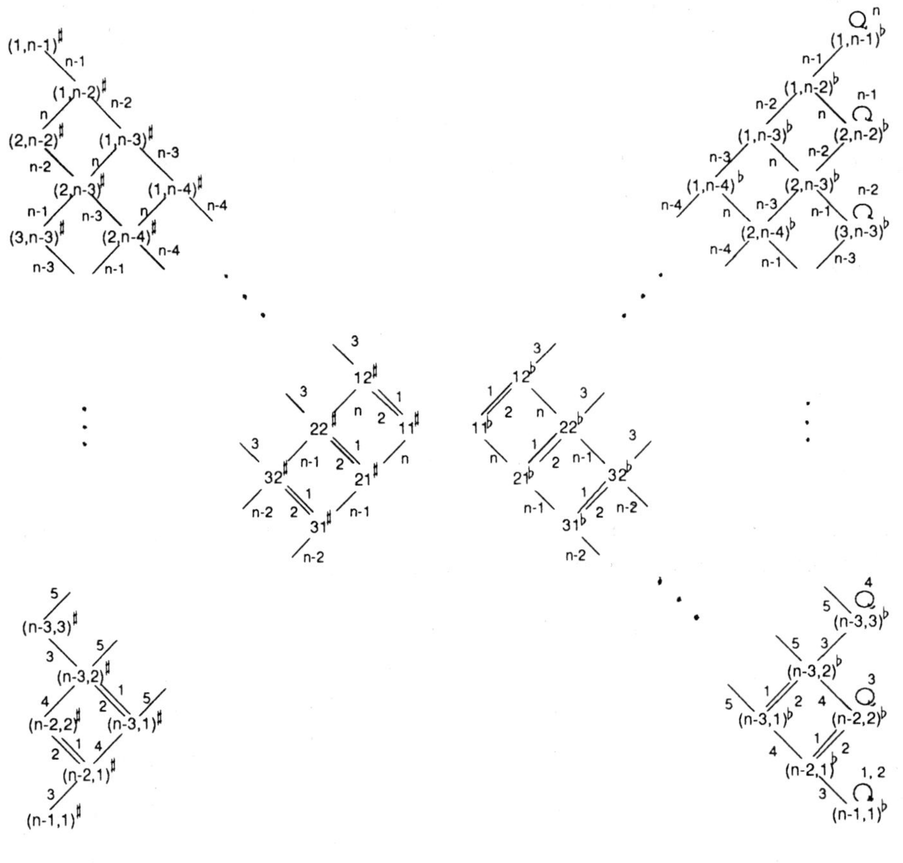

FIGURE 2.4. $\mathcal{D}_m$ for $SO_e(2, 2n-2)$

Using Table 2.2 and (2.6), $|(M_n)_{\mathrm{ds}}| = 2$. Consequently,

$$\text{(2.10g)} \qquad \begin{aligned} \mathcal{D}_{P_n} &= \mathcal{P}_{+,M_n} \cup \mathcal{P}_{-,M_n}; \\ \mathcal{P}_{\pm,M_n} &= W^{\mathbf{P}_n}_{\mathrm{upper}}. \end{aligned}$$

The pyramids in (2.10g) are the upper halves of $(D_n, A_1 \times D_{n-2})$ Hasse diagrams. Such diagrams are parametrized as the upper half of [2, Fig. 4.3]; one must relabel the Dynkin diagram of [2, (3.1)] to be compatible with (2.10a). Thus, we may parametrize the pyramids by pairs

$$\text{(2.10h)} \qquad \begin{aligned} \mathcal{P}_{+,M_n} &= \{(i,j) \mid 1 \le i \le n-1, 0 \le j \le i-1\} \\ &\quad \cup \{(i,j^*) \mid 1 \le i \le n-1, 0 \le j \le i-1\}; \\ \mathcal{P}_{-,M_n} &= \{\overline{(i,j)} \mid 1 \le i \le n-1, 0 \le j \le i-1\} \\ &\quad \cup \{\overline{(i,j^*)} \mid 1 \le i \le n-1, 0 \le j \le i-1\}. \end{aligned}$$

together with the complex type weak order relations given by the upper half of [2,Fig. 4.3]. The discrete series is parametrized by

$$\text{(2.10i)} \qquad \mathcal{D}_G = \{\hat{i} \mid 1 \le i \le 2n\}.$$

In Figure 2.5, we indicate the weak $\preceq_w$ order relations among the four subsets of (2.10c). In this figure, the order relations emanating from the bottom row are noncompact Type I. The connections from the two pyramids $\mathcal{P}_{\pm,M_n}$ to $\mathcal{T}^\sharp \subset \mathcal{D}_{P_m}$ denoted are of noncompact Type I. This gives all the weak order relations in $\mathcal{D}_o$, with the standard proviso: if a number "i" does not label an edge or loop emanating from a node δ in $\mathcal{D}$, then s_i is of compact type for the node δ.

In particular, we have shown $\mathcal{D} = \mathcal{D}_o \cup \mathcal{D}_1$. In Figure 2.6, we illustrate the situation with the $\mathcal{D}_o$ diagram associated to $SO_e(2,6)$.

(2.11) Explicit parameter set for type HS.5: In this case, $G = SO^*(2n)$. We label the Satake diagram as follows

(2.11a)

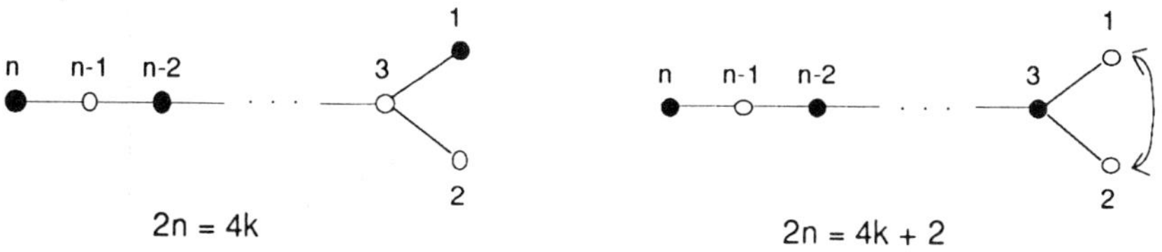

We will describe only the parametrization of $\mathcal{D}_o$; this will be sufficient for Corollary 6.22. A parametrization is implicit in [17], but we use an alternate scheme provided by Devra Garfinkle. Form the sets

$$\text{(2.11b)} \qquad \mathcal{S}(n,r)$$

consisting of arrangements δ of the numbers $1, 2, \ldots, n$ in singles or pairs subject to the following rules:

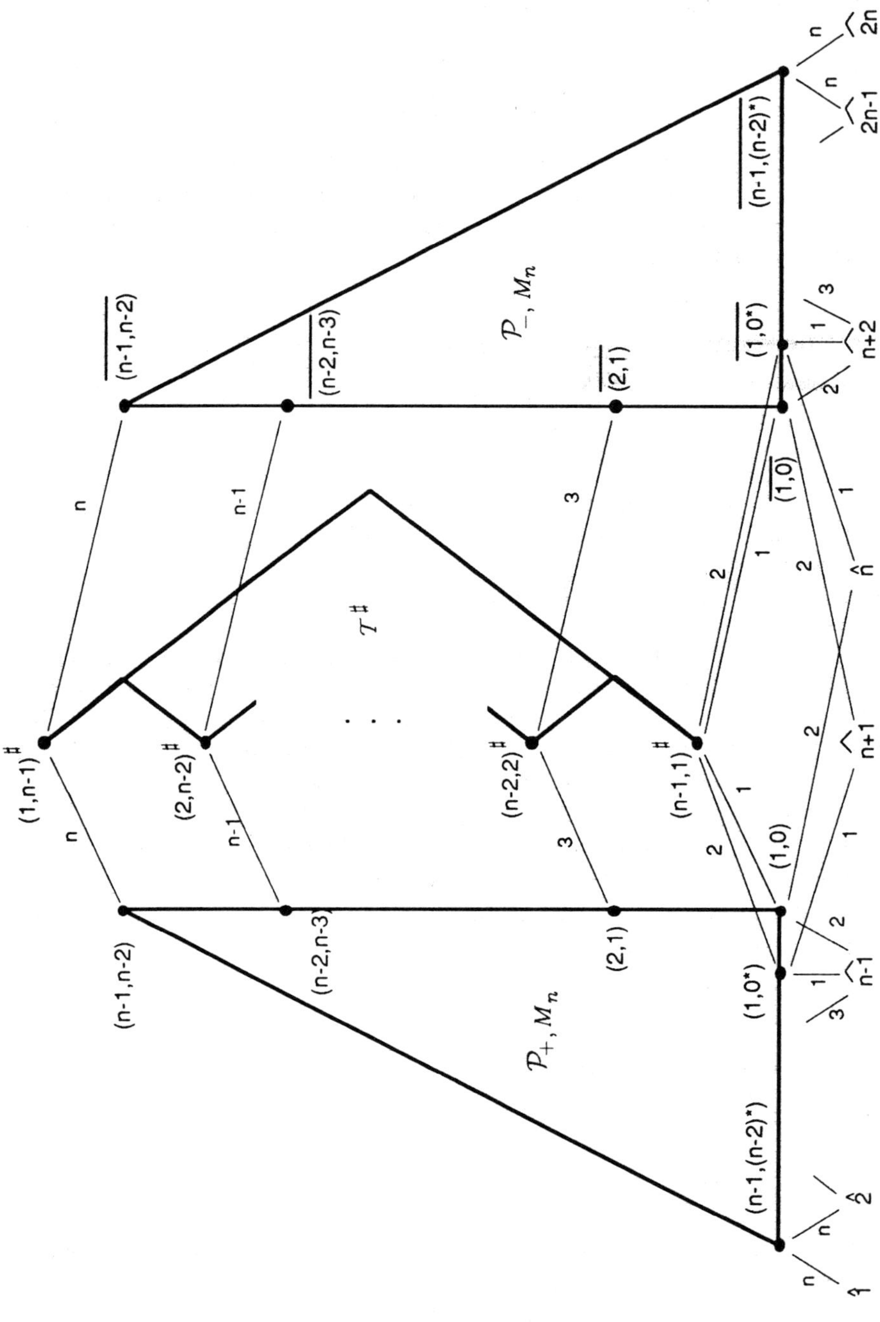

FIGURE 2.5. $\mathcal{D}_o$ for $SO_e(2, 2n-2)$

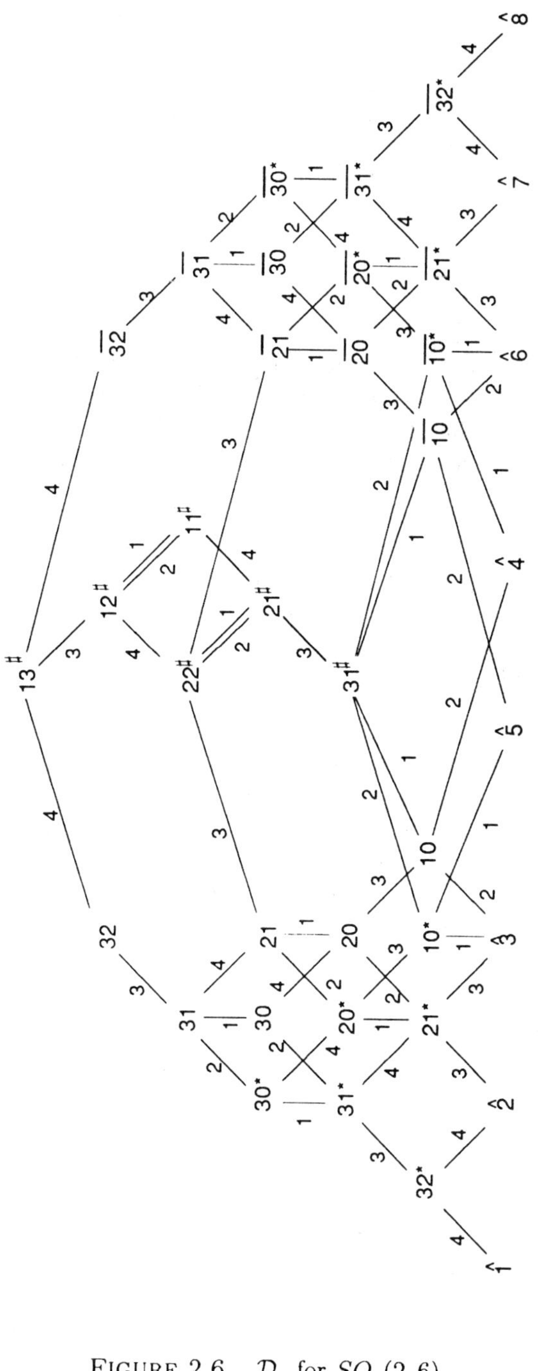

FIGURE 2.6. $\mathcal{D}_o$ for $SO_e(2,6)$

(2.11c1) Enumerate the unpaired numbers in increasing order $i_1, i_2, \ldots, i_t$ and the paired numbers $(j_1, k_1), \ldots, (j_r, k_r)$, where $j_l \leq k_l$;

(c2) Each single number i_j has a $+$ or $-$ sign attached;

(c3) Each pair (j_s, k_s) has a $\sharp$ or $\flat$ attached;

(c4) There are r pairs of numbers in the arrangement;

(c5) Let $a =$ the number of $-$ signs in δ, $b =$ the number of $\flat$'s in δ. Then we require that $a + b + r \equiv 0 \pmod 2$.

We then define

$$\mathcal{D}' = \bigcup_{r=0}^{[n/2]} S(n,r) \tag{2.11d}$$

It is an easy application of the formula (2.6c) to verify $|\mathcal{D}_o| = |\mathcal{D}'|$.

The main point is that we can make (2.11d) into a partially ordered set and with this ordering we obtain the $\preceq_G$ order of $\mathcal{D}_o$. We need to define the notions of compact, complex, etc. for reflections acting on the parameters $\mathcal{D}'$. To this end, let $\alpha_1 = e_1 + e_2$ and $\alpha_i = e_i - e_{i-1}$, for $2 \leq i \leq n$. Given $\delta \in \mathcal{D}'$, define

$$\theta_\delta(e_i) = \begin{cases} e_i & \text{if } i^\epsilon \in \delta,\ \epsilon \in \{+,-\} \\ e_j & \text{if } (i,j)^\sharp \in \delta \\ -e_j & \text{if } (i,j)^\flat \in \delta\ . \end{cases} \tag{2.11e}$$

Given θ_δ, we can follow the usual scheme (as laid out in detail in the HS.1 case) for defining compact, complex, real and noncompact roots for δ. In turn, this leads us to define a tau invariant attached to $\delta \in \mathcal{D}'$. Finally, using [21, 8.1.4] as a formal definition, we then arrive at a length function ℓ^I on $\mathcal{D}'$.

Next, we define a cross action on $\mathcal{D}'$ as follows. Let $\delta \in \mathcal{D}'$. For $2 \leq j \leq n$, $s_j \times \delta$ acts by interchanging "j" and "$j-1$". On the other hand, defining the action $s_1 \times \delta$ is a bit more involved. If δ involves the pair $(1,2)^\sharp$ or $(1,2)^\flat$, then $s_1 \times \delta = \delta$. Likewise, if δ involves 1^ϵ and $2^{-\epsilon}$, $\epsilon \in \{+,-\}$, then $s_1 \times \delta = \delta$. Otherwise, $s_1 \times \delta$ is obtained from δ by first interchanging the "1" and "2", then reversing any $\pm$ signs attached to "1" or "2", and reversing any $\sharp$ or $\flat$ attached to a pair involving "1" or "2".

Finally, we define a circle action. Let $\delta \in \mathcal{D}'$ and assume that s_j is of noncompact type for δ. Define

$$\begin{aligned} s_j \circ \delta &= (\delta \setminus \{(j-1)^\epsilon, j^{-\epsilon}\}) \cup (j-1,j)^\sharp,\ 2 \leq j \leq n; \\ s_1 \circ \delta &= (\delta \setminus \{1^\epsilon, 2^\epsilon\}) \cup (1,2)^\flat. \end{aligned} \tag{2.11f}$$

With these definitions, we have made $\mathcal{D}'$ into a poset, to which we can formally associate a Hecke module. One now needs to verify, using the theory of [24], the following result, which tells us that we may use the parameter set $\mathcal{D}'$ in future applications.

LEMMA 2.11g. *The sets $\mathcal{D}_o$ and $\mathcal{D}'$ are isomorphic as labelled posets (in the sense explained below).*

The notion of "isomorphism" in (2.11g) means that we can construct a bijection $\Psi : \mathcal{D}' \to \mathcal{D}_o$ such that the following conditions hold for all γ, $\delta \in \mathcal{D}'$:

(2.11h1) $\ell^I(\delta) = \ell^I(\Psi(\delta))$;
(h2) $\gamma \preceq_G \delta$ if and only if $\Psi(\gamma) \preceq_G \Psi(\delta)$;
(h3) the simple reflection s is of real (resp. compact, complex, etc.) type for δ if and only if it is of the same type for $\Psi(\delta)$;
(h4) if $\gamma \xrightarrow{s} \delta$, then $\Psi(\gamma) \xrightarrow{s} \Psi(\delta)$.

We should remark that $SO^*(8)$ is locally isomorphic to $SO_e(2,6)$, which we discussed in (2.10); the reader may wish to consult Figure 2.6 and convert to the parameter set described here.

(2.12) Explicit parameter sets for types HS.6, HS.7: First, suppose we are in the case of $E_{6,-14}$. We label the Satake diagram as follows:

(2.12a)

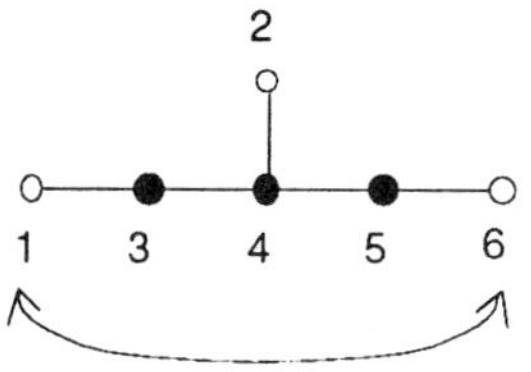

By [19], we find three conjugacy classes of Cartan subgroups:

(2.12b)
$$H_m = T_m A_m = \text{a maximally split Cartan subgroup;}$$
$$H_{16} = T_{16} A_{16} = \text{a Cartan subgroup with } \alpha_1, \alpha_6 \in \Phi_{P_{16}};$$
$$H_c = T_c A_c = \text{a compact Cartan subgroup.}$$

Define

(2.12c)
$$\mathcal{D}_{P_m}, \mathcal{D}_{P_{16}} \text{ and } \mathcal{D}_G$$

as in (2.4a). Then

(2.12d)
$$\mathcal{D} = \mathcal{D}_{P_m} \cup \mathcal{D}_{P_{16}} \cup \mathcal{D}_G.$$

We now use the theory of [24] to arrive at the data in Table 2.3; in particular, this determines the cardinality of $\mathcal{D}$ to be 513. Moreover, since H_m is connected, $\mathcal{D} = \mathcal{D}_o$.

TABLE 2.3. Cartan data for $E_{6,-14}$

H	$\dim A$	P	M_{ss}	$\lvert H/H_o \rvert$	$\lvert W(G,H) \rvert$	$\lvert \mathcal{D}_P \rvert$
H_m	2	P_m	$SU(4)$	1	$2^3 \times 4!$	270
H_{16}	1	P_{16}	$SU(5,1)$	1	$2^4 \times 3 \times 5$	216
H_c	0	G	G	1	$2^4 \times 5$	27

A further application of the theory in [24], implemented on a computer, leads us to a description of the weak $\preceq_w$ order on $\mathcal{D}$. This information is given in Table 9.1 in the Appendix.

In the case of $G = E_{7,-25}$, we have Satake diagram

(2.12f)

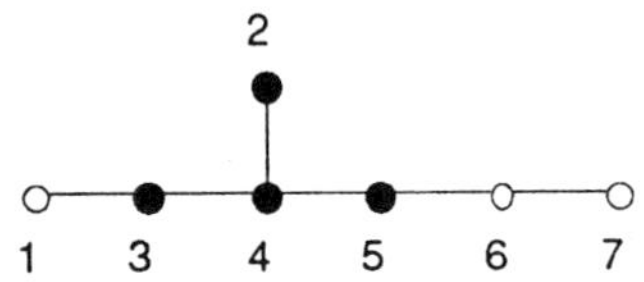

According to [19], there are four conjugacy classes of Cartan subgroups:

(2.12g)
$$\begin{aligned} H_m &= T_m A_m = \text{a maximally split Cartan subgroup;} \\ H_7 &= T_7 A_7 = \text{a Cartan subgroup with} \quad \alpha_7 \in \Phi_{P_7}; \\ H_{67} &= T_{67} A_{67} = \text{a Cartan subgroup with} \quad \alpha_6, \alpha_7 \in \Phi_{P_{67}}; \\ H_c &= T_c A_c = \text{a compact Cartan subgroup.} \end{aligned}$$

Define

(2.12h)
$$\mathcal{D}_{P_m}, \mathcal{D}_{P_7}, \mathcal{D}_{P_{67}} \text{ and } \mathcal{D}_G$$

as in (2.4a). Then

(2.12i)
$$\mathcal{D} = \mathcal{D}_{P_m} \cup \mathcal{D}_{P_7} \cup \mathcal{D}_{P_{67}} \cup \mathcal{D}_G.$$

We now use the theory of [24] to arrive at the data in Table 2.4; in particular, this determines the cardinality of $\mathcal{D}$ to be 3332. Moreover, the component group of H_m is generated by the center of $E_{7,-25}$, hence the adjoint group $PE_{7,-25}$ has a "$\mathcal{D}$ set" with data as in Table 2.4, *except* that the maximally split Cartan subgroup is now connected. The "$\mathcal{D}$ set" for the adjoint group has a single block, which will "coincide" with the fundamental block for $E_{7,-25}$. In particular, $|\mathcal{D}_o| = 3017$.

TABLE 2.4. Cartan data for $E_{7,-25}$

H	$\dim A$	P	M_{ss}	$\lvert H/H_o\rvert$	$\lvert W(G,H)\rvert$	$\lvert\mathcal{D}_{P,o}\rvert$
H_m	3	P_m	$SO(8)$	2	$2^{10} \times 3^2$	315
H_7	2	P_7	$SL_2\mathbb{R} \times SO(8)$	1	$2^9 \times 3$	1890
H_{67}	1	P_{67}	$SO_e(2,10)$	1	$2^8 \times 3 \times 5$	756
H_c	0	G	G	1	$2^7 \times 3^4 \times 5$	56

A further application of the theory in [24], implemented on a computer, leads us to a computation of the weak $\preceq_w$ order on $\mathcal{D}$. This data is given in Table 9.2 in the Appendix.

(2.13) The case of $SL_2\mathbb{R}$: In this case, there are two conjugacy classes of Cartan subgroups: a maximally split Cartan subgroup H_m, with $|H_m/H_{m,o}| = 2$ and a connected compact Cartan subgroup H_c. By general theory or explicit calculation we find that $|\mathcal{D}| = 4$. The fundamental block is described in (2.7), with $p = q = 1$, as $\mathcal{D}_o = \{\delta_{1+2-}, \delta_{1-2+}, \delta_{(12)}\}$, where δ_{1+2-} and δ_{1-2+} index two discrete series parameters and $\delta_{(12)}$ indexes the finite dimensional representation F in $\mathcal{HC}_o$. The module $\pi(\delta_{(12)})$ has the structure

(2.13a)

$\overline{\pi}(\delta_{(12)})$
$\overline{\pi}(\delta_{1+2-}) \oplus \overline{\pi}(\delta_{1-2+})$

The remaining parameter in $\mathcal{D}$, call it δ', constitutes a single block $\mathcal{D}_1 = \{\delta'\}$ and is attached to H_m. Moreover, $\pi(\delta')$ is an irreducible principal series representation induced from P_m. In summary, these remarks answer (1.1) for $SL_2\mathbb{R}$.

(2.14) A reduction: The following result will be useful in the sequel; it roughly asserts that insofar as (1.1) is concerned, we may restrict our attention to $\mathcal{D}_o$.

LEMMA 2.14a. *If G is not locally isomorphic to $Sp_n\mathbb{R}$, then $\mathcal{D}^{\max} \subset \mathcal{D}_o$.*

Proof. The exclusion of $SL_2\mathbb{R}$ is explained in (2.13). If $\mathcal{D} = \mathcal{D}_o$, then there is nothing to prove; this happens whenever G is of the following types:

(2.14b)
$$SU(p,q),\ p \neq q;$$
$$SO_e(2, 2n-1);$$
$$SO^*(4k+2);$$
$$E_{6,-14}.$$

In the remaining cases, we give a more general type argument. Because of our hypothesis, there is a unique conjugacy class of Cartan subgroup $H = TA$ with $dim_{\mathbb{R}} A = 1$, hence a unique conjugacy class of maximal cuspidal parabolic subgroup P. In these cases, H is connected. Let $\delta \in \mathcal{D}^{\max} \cap \mathcal{D}_i$ and apply successive T_s operators to obtain $\gamma \in \mathcal{D}^{\max}$ having the property that $\ell^I(\gamma) = 1$. (To see this is possible, first note that $\gamma \in \mathcal{P}_{\sigma,M,i}$, a pyramid in our block $\mathcal{D}_i$. Note that the possible posets $W^{\mathbf{P}}_{\text{upper}}$ arising are upper halves of posets of multiplicity free type considered in [2] and [7]. From these pictures it becomes clear that there is a unique upward complex type weak order relation attached to each minimal length element γ of $\mathcal{P}_{\sigma,M,i}$. By [21,(8.1.4)], $\ell^I(\gamma) = 1$.) The $\mathcal{H}$-invariance of blocks will insure that $\gamma \in \mathcal{D}_i$. Since G admits a compact Cartan subgroup, there must be a real root attached to the root system determined by γ and the connectedness of H insures the associated reflection satisfies the parity condition (see [21]). This means that $T_s \circ \gamma$ will involve a discrete series parameter. So, again by $\mathcal{H}$-invariance, $\mathcal{D}_i$ now contains a discrete series parameter. But the discrete series parameters are all in the fundamental block; hence so is δ. □

3. SOME $Sp_n\mathbb{R}$ RESULTS

In this section we prove Theorem B. Assume that $G = Sp_n\mathbb{R}$, with Satake diagram as in (2.9a). For this section only, let $\mathcal{D}(n)$ denote the set $\mathcal{D}$ for $Sp_n\mathbb{R}$, and similarly for $\mathcal{D}^{\max}(n)$, $\mathcal{D}_o(n)$, $\mathcal{D}_o^{\max}(n)$, etc. (Recall that $\mathcal{D}_o$ denotes the fundamental block of $\mathcal{D}$; see section (2.6).) The proof consists of three parts. First, we show that $\pi(\delta)$ is multiplicity free for all $\delta \in \mathcal{D}^{\max}(3)$. (In the case $n = 2$, $\mathcal{D}_o$ has been computed by D. Vogan [22].) Second, we exhibit a specific parameter $\delta \in \mathcal{D}_o^{\max}(4)$ such that $\pi(\delta)$ is not multiplicity free. Finally, we observe that there are inclusions of labelled posets $j_n : \mathcal{D}_o(n) \hookrightarrow \mathcal{D}_o(n+1)$ which are "compatible" with the associated Hecke module actions, and $j_n(\mathcal{D}_o^{\max}(n)) \subset \mathcal{D}_o^{\max}(n+1)$. In particular, for $\delta \in \mathcal{D}_o(n)$ and $\gamma' \in \mathcal{D}_o(n+1)$, $\overline{\pi}(\gamma')$ is a composition factor of $\pi(j_n(\delta))$ if and only if $\gamma' = j_n(\gamma)$ for some $\gamma \in \mathcal{D}_o(n)$, in which case

$$[\pi(j_n(\delta)) : \overline{\pi}(j_n(\gamma))] = [\pi(\delta) : \overline{\pi}(\gamma)].$$

These observations, together with the explicit computations for $n = 2, 3, 4$, imply Theorem B.

There is one complication we must deal with in the $Sp_n\mathbb{R}$ case which does not arise in any of the other groups we consider: namely, $\mathcal{D}^{\max} \not\subseteq \mathcal{D}_o$ (recall Lemma 2.14a). In fact, $\mathcal{D}^{\max}(n) \subseteq \mathcal{D}_o(n) \cup \mathcal{D}_1(n)$, where, for $0 \le i \le n$, $\mathcal{D}_i(n)$ is the block *dual* to the fundamental block of $SO(n+1+i, n-i)$, with "duality" in Vogan's sense [24]. (Caution: the groups $SO(n+1+i, n-i)$ are *disconnected* whenever $i < n$.) Put $\mathcal{D}_1^{\max}(n) = \mathcal{D}^{\max}(n) \cap \mathcal{D}_1(n)$.

We begin with the case $n = 3$. The set $\mathcal{D}_o(3)$, together with its weak order, is given in Figure 3.1. As usual, the loops denote real type reflections not satisfying the parity condition, and any simple roots omitted are of compact type. The pyramids constituting $\mathcal{D}_o^{\max}(3)$ are circled. The analogous information for $\mathcal{D}_1(3)$ is presented in Figure 3.2. The radical filtrations of the standard modules $\pi(\delta)$, $\delta \in \mathcal{D}^{\max}(3)$, were calculated by the procedure outlined in (2.5), using the recursion formulas for Lusztig-Vogan polynomials given in [22]. The results are given in Tables 3.1 and 3.2. In the tables, the number k denotes the irreducible representation $\overline{\pi}(\delta_k) = \overline{\pi}(k)$ as parametrized by Figure 3.1 or 3.2; the levels within each standard representation correspond to the semisimple layers of the radical filtration; and the representation $\pi(\delta_k) = \pi(k)$ is of course the one whose top level is k. Now it is clear by inspection that the composition factors of each standard module occur without multiplicities.

In case $n = 2$, the fundamental block $\mathcal{D}_o(2)$ is studied in [22]. In fact, as will become clearer below, it is possible to glean this poset from Figure 3.1: $\mathcal{D}_o(2)$ is

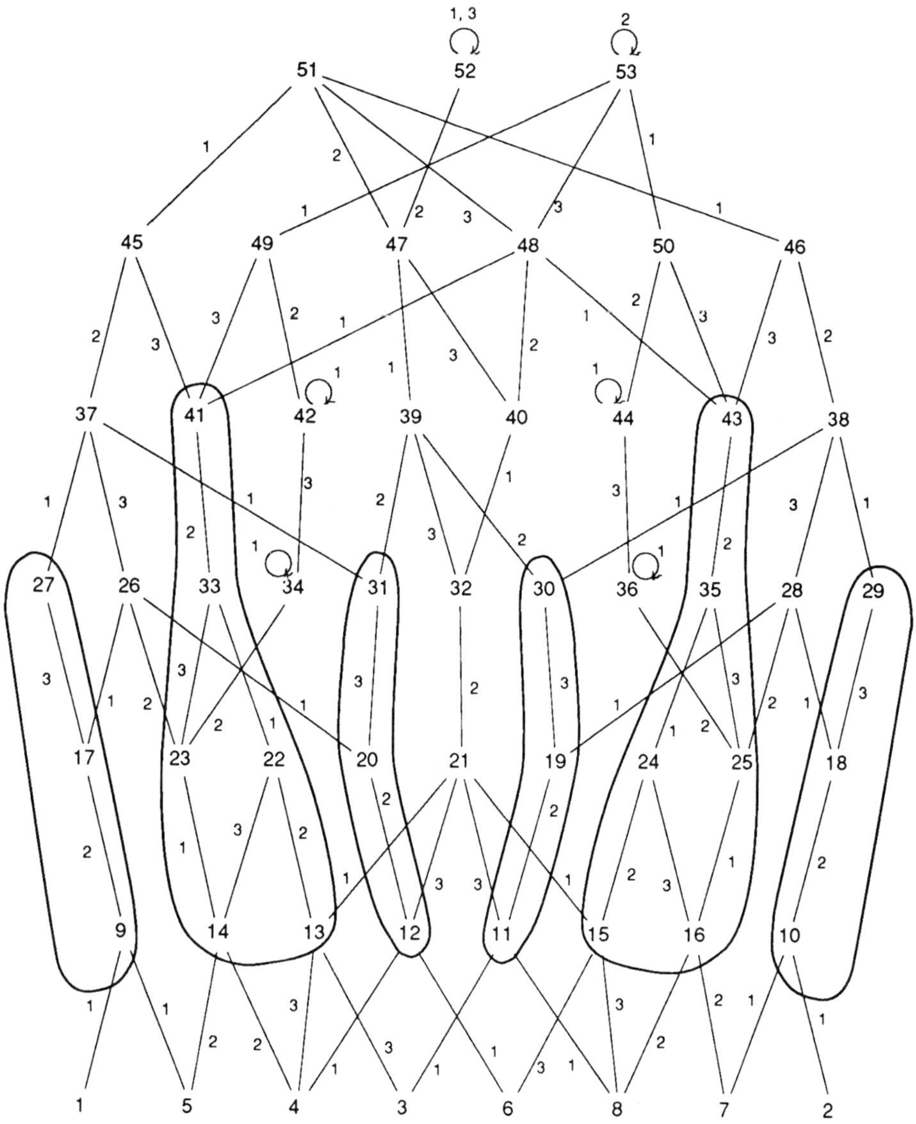

FIGURE 3.1. $\mathcal{D}_o$ for $Sp_3\mathbb{R}$

isomorphic as a labelled poset (in the sense of (2.11h)) to

$$\widetilde{\mathcal{D}}_o(3) = \{\, \delta \in \mathcal{D}_o(3) \mid \delta \preceq_G \delta_{28} \ \text{ or } \ \delta \preceq_G \delta_{36} \,\}$$

and we can read off the composition series for modules attached to $\mathcal{D}_o^{\max}(2)$ using Table 3.1. Similarly, $\mathcal{D}_1(2)$ is isomorphic to

$$\widetilde{\mathcal{D}}_1(3) = \{\, \delta \in \mathcal{D}_1(3) \mid \delta \preceq_G \delta_{13} \,\}$$

TABLE 3.1. Radical filtrations for $Sp_3\mathbb{R}$, parameters in $\mathcal{D}_o^{\max}$

9	10	11	12	13	14	15	16	17
5 1	7 2	8 3	6 4	4 3	5 4	8 6	8 7	14 9 5

18	19	20	22	23	24
16 10 7	16 11 8	14 12 4	14 13 4	14 12 9 5 4	16 15 8

25	27	29	30	31
16 11 10 8 7	22 17 14	24 18 16	24 21 19 7 16 15 11 8	22 21 20 5 14 13 12 4

33	35	41	43
23 22 21 14 13 12 4	25 24 21 16 15 11 8	34 33 32 26 23 22 21 20 6 3 14 13 12 4	36 35 32 28 25 24 21 19 6 3 16 15 11 8

and the composition series for modules attached to $\mathcal{D}_1^{\max}(2)$ follow from Table 3.2.

We next turn to the case $n = 4$. In Figure 3.3 we give a portion of the labelled poset $\mathcal{D}_o(4)$, including one complete pyramid, which is circled. Since only a portion of the diagram is shown, we use the new notation

$$\overset{k}{\underset{i}{\circlearrowleft}}$$

to indicate that the simple root α_i is of compact type for the parameter δ_k. (Do not confuse this with the notation for a "real root not in the tau invariant," depicted by a "loop above the parameter," as in Figure 3.1.) Enough data is given so that the following can be computed, using the U_α-algorithm (see [22, Proposition 8.13] and Chapter 6):

(i) The complete radical filtrations of $\pi(34)$, $\pi(48)$, $\pi(80)$, and $\pi(116)$;
(ii) $[\pi(152) : \overline{\pi}(81)] = 1$;
(iii) $[\pi(152) : \overline{\pi}(117)] = [\pi(152) : \overline{\pi}(116)] = 1$, $[\pi(152) : \overline{\pi}(61)] \geq 1$.

It follows from (iii) that $[\Theta_s(\pi(152)) : \overline{\pi}(81)] \geq 3$, where $s = s_3$. Hence (ii) implies that $[\pi(186) : \overline{\pi}(81)] \geq 2$. Since $186 \in \mathcal{D}_o^{\max}$, this gives the desired contradiction to Vogan's conjecture.

Finally, we describe the inclusions $j_n : \mathcal{D}_o(n) \hookrightarrow \mathcal{D}_o(n+1)$. For this, we need an explicit parametrization of $\mathcal{D}_o(n)$. Such a parametrization has been described by D. Garfinkle (private communication), and can be formulated as follows. Each element of $\mathcal{D}_o(n)$ corresponds to an arrangement δ of the numbers $0, 1, 2, \ldots, n$. The numbers $1, \ldots, n$ occur singly or in pairs; the number 0 is always a singleton. Each pair has either a $+1$ or a -1 attached to it. The singleton 0 has either a $+$ sign or a $-$ sign attached to it. Each remaining singleton has one of the

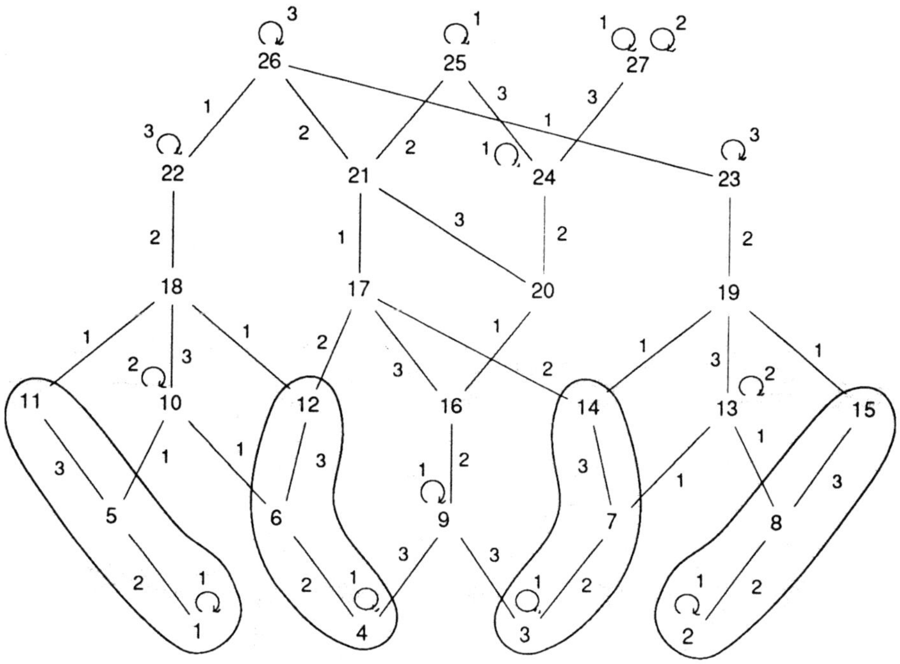

FIGURE 3.2. $\mathcal{D}_1$ for $Sp_3\mathbb{R}$

four symbols $+$, $-$, $\sharp$, or $\flat$ attached to it. Let $r(\delta)$ be the number of pairs, $s(\delta)$ the total number of $\sharp$'s and $\flat$'s, $t(\delta)$ the number $+$ signs attached to singletons among $1, \ldots, n$, and $u(\delta)$ the number of $-$ signs attached to singletons among $1, \ldots, n$. Finally, define

$$\epsilon^+(\delta) = \begin{cases} 1 & \text{if } 0^+ \text{ occurs in } \delta \\ 0 & \text{if } 0 \text{ occurs in } \delta. \end{cases}$$

Then we require further that

$$t(\delta) + \epsilon^+(\delta) = u(\delta) + 1. \tag{3.1}$$

If $\pi(\delta)$ is induced from the parabolic subgroup $P_\delta = M_\delta A_\delta N_\delta$, then

$$\dim_{\mathbb{R}} A_\delta = r(\delta) + t(\delta) + u(\delta).$$

The action of the simple root α_i on δ is determined by the status of the numbers $i-1$ and i in δ, and only these two coordinates are affected by s_i.

TABLE 3.2. Radical filtrations for $Sp_3\mathbb{R}$, parameters in $\mathcal{D}_1^{\max}$

1	2	3	4	5	6	7	8	11	12	14	15
				1	4	3	2	5	9 6	9 7	8
									4	3	

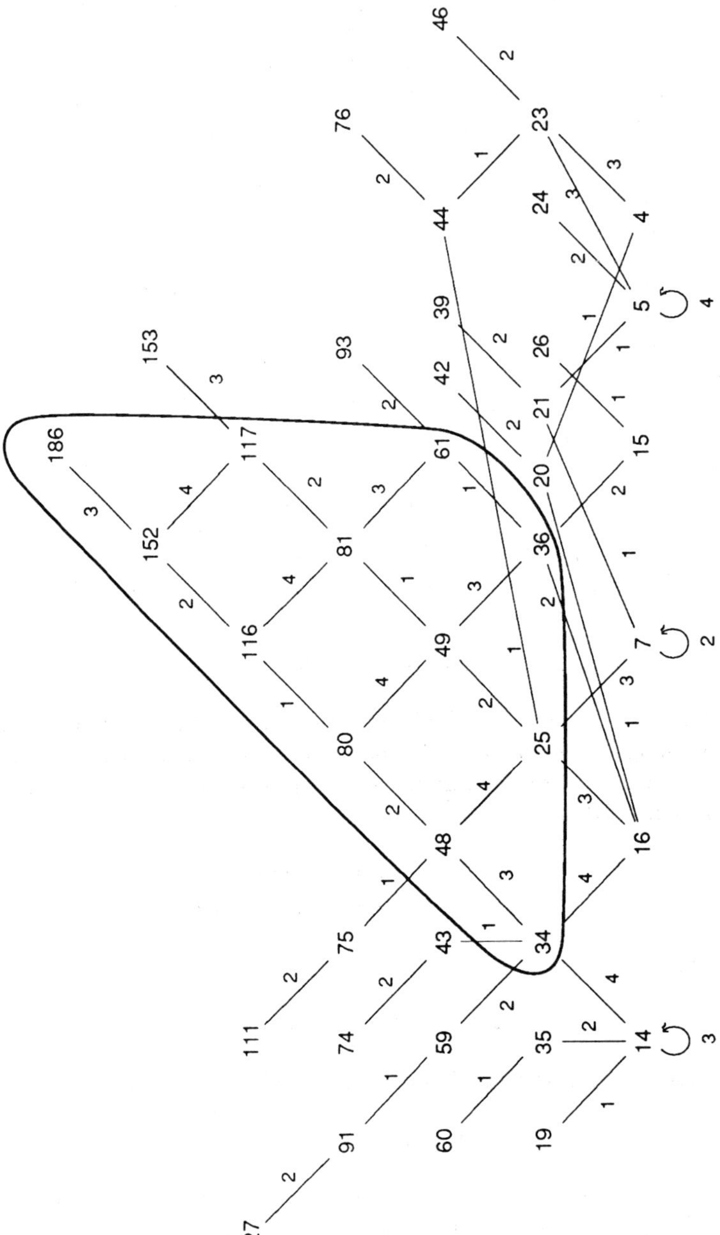

FIGURE 3.3. Part of $\mathcal{D}$ for $Sp_4\mathbb{R}$

Given such a parameter $\delta \in \mathcal{D}_o(n)$, define a new parameter $\delta' = j_n(\delta)$ (involving $0, 1, \ldots, n+1$) by appending the symbol $(n+1)^\sharp$ to δ. Then $r(\delta') = r(\delta)$, $s(\delta') = s(\delta) + 1$, $t(\delta') = t(\delta)$, $u(\delta') = u(\delta)$, and $\epsilon^+(\delta') = \epsilon^+(\delta)$. Thus δ' satisfies (3.1); i.e., $\delta' \in \mathcal{D}_o(n+1)$. Notice also that $\dim_{\mathbb{B}} A_{\delta'} = \dim_{\mathbb{B}} A_\delta$. Hence $j_n : \mathcal{D}_o(n) \to \mathcal{D}_o(n+1)$, and $j_n(\mathcal{D}_o^{\max}(n)) \subset \mathcal{D}_o^{\max}(n+1)$. In fact, it is easy to see that j_n is a bijection of $\mathcal{D}_o(n)$ with the subset $\mathcal{D}_o^\sharp(n+1)$ of $\mathcal{D}_o(n+1)$ consisting of those parameters δ' in which $(n+1)^\sharp$ occurs. Furthermore, since the coordinates $0, 1, \ldots, n$ are identical in δ and $j_n(\delta)$, the bijection between $\mathcal{D}_o(n)$ and $\mathcal{D}_o^\sharp(n+1)$ carries the action of $s_1, \ldots, s_n$ in $\mathcal{D}_o(n)$ to the corresponding action in $\mathcal{D}_o^\sharp(n+1)$. Therefore the radical filtration of each $\pi(\delta')$, $\delta' = j_n(\delta)$, is computed by the same algorithm as that of $\pi(\delta)$, if we identify corresponding parameters in $\mathcal{D}_o(n)$ and $\mathcal{D}_o^\sharp(n+1)$.

These remarks complete the proof of Theorem B.

4. INDUCING FROM HOLOMORPHIC DISCRETE SERIES

In this chapter, we realize the goal of (1.3)–(1.5) for generalized principal series representations induced from holomorphic data. The main results in this section are Theorem 4.8 and Theorem D.

THEOREM D. *Let G be simple of Hermitian type and $P{=}MAN$ a maximal cuspidal parabolic subgroup of G where the noncompact factor of M_{ss} is both simple and of Hermitian type. If σ is a holomorphic discrete series representation of M, then $\mathrm{Ind}_P^G(\sigma \otimes \nu \otimes 1)$ is multiplicity free.*

For the groups $SU(p,q)$, $SO_e(2,2n-2)$, $SO^*(2n)$, $E_{6,-14}$ and $E_{7,-25}$, every maximal cuspidal parabolic subgroup will satisfy the desired Levi factor condition of Theorem D. In the case of $SO_e(2,2n-1)$, there exists a maximal cuspidal parabolic subgroup $P = MAN$, where $M_{ss} = SO(1,2n-2)$; in particular, the Levi factor is not of Hermitian type. In the case of $Sp_n\mathbb{R}$, there are two types of maximal cuspidal parabolic subgroups: $P_1 = M_1A_1N_1$, with $M_{1,ss} = Sp_{n-1}\mathbb{R}$; $P_2 = M_2A_2N_2$, with $M_{2,ss}$ locally isomorphic to $SL_2^{\pm}\mathbb{R} \times Sp_{n-2}\mathbb{R}$. Only P_1 would be allowed in Theorem D for $G = Sp_n\mathbb{R}$.

Recalling our fixed choice of complex structure on G/K, via $\mathbf{Q}^\sharp$, a class of irreducible highest weight representations for G lying in $\mathcal{HC}_o$ is determined. These will be irreducible quotients of algebraically induced generalized Verma modules of the form $N_w = \mathfrak{U}(\mathfrak{g}) \otimes_{\mathfrak{U}(\mathfrak{q}^\sharp)} E_w$, indexed by $w \in W^{\mathbf{Q}^\sharp}$. Note that such induced modules are always $\mathfrak{g}$-modules and modules for the Levi factor $\mathfrak{k}$ of the inducing parabolic $\mathfrak{q}^\sharp$. We see N_w is a $(\mathfrak{g},\mathfrak{k})$-module with finite $\mathfrak{k}$ multiplicities, hence in $\mathcal{HC}_o$; see [13]. In view of these remarks, $\mathbf{Q}^\sharp$ determines a subset $\mathcal{D}_{\mathsf{hw}}$ of $\mathcal{D}$. In this way, we can speak of the irreducible highest weight representations $\overline{\pi}(\delta(w))$, $w \in W^{\mathbf{Q}^\sharp}$. If we choose the opposite complex structure on G/K, then a family of *lowest* weight representations will be determined, corresponding to the parabolic subgroup $\overline{\mathbf{Q}^\sharp}$ opposite to $\mathbf{Q}^\sharp$ and leading to a subset $\mathcal{D}_{\mathsf{lw}}$ of $\mathcal{D}$.

In [1] and [11] it is shown that the modules N_w are multiplicity free; this can be interpreted as saying that the standard induced modules in the category $\mathcal{O}(\mathfrak{g},\mathfrak{q}^\sharp)$ are multiplicity free. A corollary to Theorem D answers the analogous question in the Harish-Chandra category.

COROLLARY 4.1. *Let G be simple of Hermitian type. If $\delta \in \mathcal{D}_{\mathsf{hw}} \cap \mathcal{D}^{\max}$, then $\pi(\delta)$ is multiplicity free.*

Proof. One needs to observe that the Langlands data of any such $\overline{\pi}(\delta)$ involves a holomorphic discrete series and a parabolic subgroup of the type in Theorem D. This is discussed in [25] and also follows from [6]. □

If $\delta \in \mathcal{D}_{\mathrm{hw}}$, it is very important to emphasize that although the Langlands quotient of $\pi(\delta)$ is a highest weight module, not all of the composition factors need be of highest weight type; recall (1.7). For example, in the case $G=SU(2,1)$, with the notation of [7,(5.3.1)], the standard module denoted $\pi_{0,2}(\chi)$ contains a large discrete series π_1 as a subquotient and π_1 is not a highest weight module for any complex structure on G/K. Our proof of Theorem D will proceed by giving closed formulas for the radical filtration of standard modules in the so-called "holomorphic pyramid". To define these special pyramids, we fix a maximal cuspidal parabolic subgroup $P = MAN$ having the property that the non-compact factor of M_{ss} is simple and of Hermitian symmetric type. Let σ be the class of a holomorphic discrete series representation for M and denote by $\mathcal{P}_{\mathrm{holo},M}$ the pyramid $\mathcal{P}_{\sigma,M}$; we refer to this as a *holomorphic pyramid* in $\mathcal{D}$.

Our formulas depend on the theory of Enright-Shelton [11] and the orbit decomposition analysis in [9]. Consequently, we begin with the relevent terminology. As noted in the introduction, a fixed choice of complex structure on G/K determines a parabolic subgroup $\mathbf{Q}^\sharp$ of $\mathbf{G}$; but, keep in mind that $\mathbf{Q}^\sharp$ is not defined over $\mathbb{R}$. The Levi decomposition of $\mathbf{Q}^\sharp$ over $\mathbb{C}$ has the form $\mathbf{Q}^\sharp = \mathbf{KU}$ and as a consequence, every $\mathbf{Q}^\sharp$-orbit is $\mathbf{K}$-stable and itself a union of $\mathbf{K}$-orbits. This allows us to define maps $\mathbb{A}, \mathbb{B}$ and $\mathbb{B}^i$, $i \in \mathbb{N}$, relating the set V (which indexes the $\mathbf{K}$-orbits) and the set $W^{\mathbf{Q}^\sharp}$ (which indexes the $\mathbf{Q}^\sharp$-orbits) as follows. First, since $\mathbf{K}$ is connected, each $\mathbf{Q}^\sharp$-orbit $\mathcal{O}_w$ contains a unique open dense $\mathbf{K}$-orbit, denoted $\mathcal{V}_{\mathbb{B}(w)}$, $\mathbb{B}(w) \in V$. This allows us to define

$$\mathbb{B} : W^{\mathbf{Q}^\sharp} \longmapsto V. \tag{4.2}$$

Next, define

$$\mathbb{A} : V \longmapsto W^{\mathbf{Q}^\sharp}; \tag{4.3}$$

by $\mathbb{A}(\delta_o) = w$, where $\mathcal{V}_{\delta_o} \subset \mathcal{O}_w$. The above remarks show $\mathbb{A}$ is well defined and the next result gives an effective means of computing it. For each fixed $w \in W^{\mathbf{Q}^\sharp}$, define

$$S_w = \{\, s \in S \mid ws < w \,\}.$$

PROPOSITION 4.4 [9, (1.2)]. *If $w \in W^{\mathbf{Q}^\sharp}$, then we have*

$$\mathcal{O}_w = \bigcup_{\gamma_o \in V_w} \mathcal{V}_{\gamma_o}$$

where

$$V_w = \{\, \gamma_o \in V \mid \gamma_o \preceq \mathbb{B}(w) \quad \text{and} \quad \gamma_o \npreceq \mathbb{B}(ws), \quad \text{for all} \quad s \in S_w \,\}.$$

We let

$$\begin{aligned} &\jmath : V \longmapsto \mathbb{N}; \\ &\jmath(\delta_o) = \text{codimension of } \mathcal{V}_{\delta_o} \text{ in } \mathcal{O}_{\mathbb{A}(\delta_o)}, \end{aligned} \tag{4.5}$$

which gives rise to the family of *set valued mappings*

$$\mathbb{B}^i : W^{\mathbf{Q}^\sharp} \longmapsto \{\text{subsets of } V\}; \tag{4.6}$$
$$\mathbb{B}^i(w) = \{\, \delta_o \in V \mid \mathcal{V}_{\delta_o} \subset \mathcal{O}_w \text{ and } \jmath(\delta_o) = i\}.$$

Observe, the computation of $\jmath$ and $\mathbb{B}^i$ depends on (4.4) and also, $\mathbb{B} = \mathbb{B}^0$. Note that the maps $\mathbb{B}^i$ are in general set-valued, but in the context of our next theorem, $\mathbb{B}^i(y)$ is always either the empty set or a subset of V consisting of a single element.

For each $w \in W^{\mathbf{Q}^\sharp}$, we denote by $\mathcal{E}_w$ the collection of sets of orthogonal roots associated to w by Enright-Shelton in [11, (8.4)]. Denote by $\Delta(\mathfrak{u})$ the roots of $\mathfrak{h}$ in $\mathfrak{u}$ and by $\Phi^+_{\mathbf{Q}^\sharp}$ the positive roots for the Levi factor of $\mathfrak{q}^\sharp$ determined by the Borel subalgebra $\mathfrak{b}$. We make two auxiliary definitions. Given $\Omega \in \mathcal{E}_w$, we define $\Omega^+ = \{\beta \in \Omega : w\beta \in \Delta(\mathfrak{u})\}$. Also, to any such Ω we can attach the Weyl group element

$$r_\Omega = \prod_{\beta \in \Omega^+} s_\beta \tag{4.7a}$$

Observe that since the roots in Ω are mutually orthogonal the order of the product in (4.7a) does not matter. Finally, if $\lambda \in \mathfrak{h}^*$, let $\bar{\lambda}$ denote the unique $\Phi^+_{\mathbf{Q}^\sharp}$ dominant element in the $W_{\mathbf{Q}^\sharp}$-orbit of λ.

Let $\delta \in \mathcal{P}_{\text{holo},M}$ and consider the expression

$$\mathbb{B}^i\left(\overline{\mathbb{A}(\delta_o) r_\Omega}\right), \ \Omega \in \mathcal{E}_{\mathbb{A}(\delta_o)} \tag{4.7b}$$

The term in (4.7b) is the *set* of $\mathbf{K}$-orbit parameters in V indexing the $\mathbf{K}$-orbits of codimension i inside the $\mathbf{Q}^\sharp$-orbit $\mathcal{O}_{\overline{\mathbb{A}(\delta_o) r_\Omega}}$; possibly even the empty set.

We now state the main theorem of this section, which answers (1.3)–(1.5) for parameters in the holomorphic pyramids; recall, parameters in such a pyramid are attached (by definition) to a parabolic subgroup $P = MAN$ satisfying the hypotheses of Theorem D.

THEOREM 4.8. *Assume G is simple of Hermitian type and $\delta \in \mathcal{P}_{\text{holo},M}$. Let $\mathcal{E}_\delta$ denote the Enright-Shelton collection $\mathcal{E}_{\mathbb{A}(\delta_o)}$.*

(i) If $\delta = (\mathcal{V}_{\delta_o}, \mathcal{L}_{\delta_o})$, $\gamma = (\mathcal{V}_{\gamma_o}, \mathcal{L}_{\gamma_o}) \in \widehat{\mathcal{P}}_{\text{holo},M}$ and $\mathcal{V}_{\delta_o} = \mathcal{V}_{\gamma_o}$, then $\mathcal{L}_{\delta_o} = \mathcal{L}_{\gamma_o}$. In particular, we may use the notation $\hat{C}_{\gamma_o}$ to unambiguously denote the self-dual element of $\mathcal{M}_{\mathbf{K}}$ attached to the unique parameter $\gamma = (\mathcal{V}_{\gamma_o}, \mathcal{L}_{\gamma_o})$ in $\widehat{\mathcal{P}}_{\text{holo},M}$.

(ii) For every $\Omega \in \mathcal{E}_\delta$ and $i \in \{\jmath(\delta_o), \jmath(\delta_o)+1\}$, $\mathbb{B}^i(\overline{\mathbb{A}(\delta_o) r_\Omega})$ is either a singleton set or the empty set.

(iii) Define $\mathbb{E}_\delta : \mathcal{E}_\delta \longmapsto \mathcal{M}_{\mathbf{K}}$ by

$$\mathbb{E}_\delta(\Omega) = \hat{C}_{\mathbb{B}^{\jmath(\delta_o)}\left(\overline{\mathbb{A}(\delta_o) r_\Omega}\right)} + u^{-1/2} \hat{C}_{\mathbb{B}^{\jmath(\delta_o)+1}\left(\overline{\mathbb{A}(\delta_o) r_\Omega}\right)},$$

with the proviso that we ignore the terms where $\mathbb{B}^i(\dots) = \varnothing$. Then we may express δ in $\mathcal{M}_{\mathbf{K}}$ as

$$\delta = \sum_{\Omega \in \mathcal{E}_\delta} u^{-|\Omega^+|/2} \mathbb{E}_\delta(\Omega).$$

In particular, this implies $\overline{\pi}(\gamma)$ *is a composition factor in the* $|\Omega^+| + i - \jmath(\delta_o)$ *layer of the radical filtration of* $\pi(\delta)$ *if and only if* $\gamma \in \mathbb{B}^i(\overline{\mathbb{A}(\delta_o) r_\Omega}) \neq \varnothing$ *for some* $\Omega \in \mathcal{E}_\delta$ *and* $i \in \{\jmath(\delta_o), \jmath(\delta_o)+1\}$.

(iv) The standard module $\pi(\delta)$ *is multiplicity free.*

Before beginning the proof of Theorem 4.8, it is important to emphasize that a holomorphic pyramid $\mathcal{P}_{\mathrm{holo},M}$ is typically *not* a subset of $\mathcal{D}_{\mathrm{hw}}$. If $\delta \in \mathcal{P}_{\mathrm{holo},M} \cap \mathcal{D}_{\mathrm{hw}}$, then $\mathcal{V}_{\delta_o}$ is an open dense subvariety of $\mathcal{O}_{\mathbb{A}(\delta_o)}$. But, in general, we see that the holomorphic pyramid parameters will index $\mathbf{K}$-orbits which have positive codimension in some $\mathbf{Q}^\sharp$-orbit. For example, this can already be seen in $SU(2,2)$: using the notation in [9, Table 2], $\mathcal{P}_{\mathrm{holo},M} = \{\delta_i : i = 20, 17, 16, 12, 11, 9\}$, $\mathcal{P}_{\mathrm{holo},M} \cap \mathcal{D}_{\mathrm{hw}} = \{\delta_j : j = 20, 17, 16, 11\}$; in particular, $\mathcal{V}_{\delta_{9_o}}$ has codimension one in the $\mathbf{Q}^\sharp$-orbit $\mathcal{O}_{s_2 s_3}$. In view of these remarks, our starting point (and the key to our entire enterprise) is to understand the Proposition 4.4 decomposition for those $\mathbf{Q}^\sharp$-orbits containing a $\mathbf{K}$-orbit $\mathcal{V}_{\delta_o}$, with $\delta \in \mathcal{P}_{\mathrm{holo},M}$. The approach below depends primarily upon the Hecke module formalism and underlying geometric considerations reviewed in chapter 2.

Recalling (2.5i) and (2.5j), the rather surprising content of Theorem 4.8 is that we may express the $S_{\gamma,\delta}$'s in terms of the Enright-Shelton data which computes $S^\natural_{y,w}$'s in [10]. Referring to [10, (1.3), (1.7)], we find that

$$\mathbb{I}_{\mathcal{O}_w} = \sum_{\Omega \in \mathcal{E}_w} u^{-|\Omega^+|/2} [IC(\mathcal{O}_{\overline{w r_\Omega}})] \tag{4.9}$$

In otherwords, $S^\natural_{y,w} = 0$ unless $y = \overline{w r_\Omega}$ for some $\Omega \in \mathcal{E}_w$, in which case $S^\natural_{y,w} = u^{-|\Omega^+|/2}$. Describing $S_{\gamma,\delta}$ and hence proving Theorem 4.8 will require rather explicit Proposition 4.4 type data. This data is provided in (4.18)-(4.23) below. It is important to emphasize that this data is easily obtained from the weak order on the set $\mathcal{D}$ and is not at all recursive; this is in contrast to Lusztig-Vogan polynomial calculations.

Let $\delta \in \mathcal{P}_{\mathrm{holo},M}$ and define

$$X(\delta_o, i) = \{\gamma_o \preceq \delta_o \mid \mathbb{A}(\gamma_o) = \mathbb{A}(\delta_o) \quad \text{and} \quad \jmath(\gamma_o) \geq i\} \subset V \tag{4.10}$$

This set indexes the set of $\mathbf{K}$-orbits in the intersection $\bar{\mathcal{V}}_{\delta_o} \cap \mathcal{O}_{\mathbb{A}(\delta_o)}$ which have codimension at least i in $\mathcal{O}_{\mathbb{A}(\delta_o)}$. For example, consider the setting of Corollary 4.1, so that $\delta \in \mathcal{P}_{\mathrm{holo},M} \cap \mathcal{D}_{\mathrm{hw}}$, then $\delta = \delta(w)$, $w = \mathbb{A}(\delta_o) \in W^{\mathbf{Q}^\sharp}$, $\mathcal{V}_{\delta_o}$ is dense in $\mathcal{O}_w$ and

$$X(\delta_o, 0) = \{\gamma_o \preceq \delta_o \mid \mathcal{V}_{\gamma_o} \subset \mathcal{O}_w\},$$

which is just the set of $\mathbf{K}$-orbits in $\mathcal{O}_w$. If $\delta \in \mathcal{P}_{\mathrm{holo},M}$, we define

$$\mathcal{U}^i_\delta = \bigcup_{\gamma_o \in X(\delta_o, i)} \mathcal{V}_{\gamma_o} \tag{4.11}$$

The next lemma is the key result needed to prove Theorem 4.8 and follows from (4.18)-(4.23) below.

LEMMA 4.12. *Let $\delta \in \mathcal{P}_{\mathrm{holo},M}$, then we have*

(i) For all $y \leq \mathbb{A}(\delta_o)$, $\mathbb{B}^{\jmath(\delta_o)}(y)$ and $\mathbb{B}^{\jmath(\delta_o)+1}(y)$ are either empty or singelton sets.

(ii) If $i = \jmath(\delta_o)$ or $\jmath(\delta_o)+1$, then

$$\overline{\mathcal{U}}_\delta^i = \bigcup_{y \leq \mathbb{A}(\delta_o), \mathbb{B}^i(y) \neq \emptyset} \mathcal{U}^i_{\mathbb{B}^i(y)}.$$

(iii) If we stratify the variety $\overline{\mathcal{U}}_\delta^{\jmath(\delta_o)}$ as in (ii), we obtain the analog of [4,(3.12)] where $\mathcal{O}_w$ is replaced by $\mathcal{U}_\delta^{\jmath(\delta_o)}$. A similar remark holds for the variety $\overline{\mathcal{U}}_\delta^{\jmath(\delta_o)+1}$.

Remark 4.13. It is worth emphasizing that the content of (4.12)(iii) is that we may compute the Euler characteristic of $IC(\mathcal{U}_\delta^{\jmath(\delta_o)})$ using the Hecke module structure of $\mathcal{M}_{\mathbf{Q}^\sharp}$, as described in [4].

Proof of Theorem 4.8. In the Grothendieck group $K(\mathcal{E}(\mathbf{K})')$, (4.12)(i) implies

$$\mathbb{I}_{\mathcal{U}_\delta^{\jmath(\delta_o)}} = \mathbb{I}_{\mathcal{V}_{\delta_o}} + \mathbb{I}_{\mathcal{U}_\delta^{\jmath(\delta_o)+1}},$$

hence

$$\mathbb{I}_{\mathcal{V}_{\delta_o}} = \mathbb{I}_{\mathcal{U}_\delta^{\jmath(\delta_o)}} - \mathbb{I}_{\mathcal{U}_\delta^{\jmath(\delta_o)+1}}. \tag{4.14}$$

Since $\mathcal{V}_{\delta_o}$ is dense in $\mathcal{U}_\delta^{\jmath(\delta_o)}$, $IC(\mathcal{U}_\delta^{\jmath(\delta_o)}) = IC(\mathcal{V}_{\delta_o})$ and (4.13) together with (4.9) imply the existence of Laurent polynomials $T_{y,z}$ such that

$$\begin{aligned}\mathbb{I}_{\mathcal{U}_\delta^i} &= \sum_{y \leq \mathbb{A}(\delta_o)} T_{y,\mathbb{A}(\delta_o)}(u)[IC(\mathcal{U}^i_{\mathbb{B}^i(y)})] \\ &= \sum_{\Omega \in \mathcal{E}_{\mathbb{A}(\delta_o)}} u^{(-|\Omega^+|+\jmath(\delta_o)-i)/2}[IC(\mathcal{U}^i_{\mathbb{B}^i(\overline{\mathbb{A}(\delta_o)r_\Omega})})]\end{aligned} \tag{4.15}$$

for $i = \jmath(\delta_o)$ or $\jmath(\delta_o)+1$. Combining (4.14) and (4.15) and taking into account a Tate twist, which enters into the picture when viewing $\bar{\mathcal{U}}_\delta^{\jmath(\delta_o)+1}$ as a subvariety of codimension one in $\bar{\mathcal{U}}_\delta^{\jmath(\delta_o)}$, we get

$$\begin{aligned}\mathbb{I}_{\mathcal{V}_{\delta_o}} &= \sum_{\Omega \in \mathcal{E}_{\mathbb{A}(\delta_o)}} u^{-|\Omega^+|/2}[IC(\mathcal{U}^{\jmath(\delta_o)}_{\mathbb{B}^{\jmath(\delta_o)}(\overline{\mathbb{A}(\delta_o)r_\Omega})})] \\ &\quad - \sum_{\Omega \in \mathcal{E}_{\mathbb{A}(\delta_o)}} u^{(-|\Omega^+|-1)/2}[IC(\mathcal{U}^{\jmath(\delta_o)+1}_{\mathbb{B}^{\jmath(\delta_o)+1}(\overline{\mathbb{A}(\delta_o)r_\Omega})})] \\ &= \sum_{\Omega \in \mathcal{E}_{\mathbb{A}(\delta_o)}} u^{-|\Omega^+|/2}[IC(\mathcal{V}_{\mathbb{B}^{\jmath(\delta_o)}(\overline{\mathbb{A}(\delta_o)r_\Omega})})] \\ &\quad - \sum_{\Omega \in \mathcal{E}_{\mathbb{A}(\delta_o)}} u^{(-|\Omega^+|-1)/2}[IC(\mathcal{V}_{\mathbb{B}^{\jmath(\delta_o)+1}(\overline{\mathbb{A}(\delta_o)r_\Omega})})].\end{aligned} \tag{4.16}$$

But, if $\delta \in \mathcal{P}_{\mathrm{holo},M}$, (4.16) computes the weight filtration of $\pi(\delta)$, which coincides with the radical filtration (by Casian's result [3]); this proves Theorem 4.8.

It remains to present the technical results necessary to verify Lemma 4.12. We proceed, case-by-case, using the indexing HS.1-HS.7 in chapter 2.

(4.17) Strategy: We will be applying the algorithm of [9] to decompose $\mathbf{Q}^\sharp$-orbits into $\mathbf{K}$-orbits. We will express these decompositions as formal elements in the module

$$\mathcal{M}_{\mathrm{orb}} = \mathbb{Z}[t] \otimes_{\mathbb{Z}} \mathbb{Z}[V]$$

as follows: Recall the notation V_w in (4.4). We may view each $w \in W^{\mathbf{Q}^\sharp}$ as an element of $\mathbb{Z}[V]$; given the decompostion $\mathcal{O}_w = \bigcup_{\gamma_o \in V_w} \mathcal{V}_{\gamma_o}$, we let $w = \Sigma_{\gamma_o \in V_w} \gamma_o \in \mathbb{Z}[V]$. In other words, V and $W^{\mathbf{Q}^\sharp}$ are both viewed as bases for $\mathbb{Z}[V]$. Since $\mathcal{V}_{\gamma_o}$ has codimension $\jmath(\gamma_o)$ in $\mathcal{O}_w$, we will write the element

$$w = \Sigma_{\gamma_o \in V_w} t^{\jmath(\gamma_o)} \gamma_o \in \mathcal{M}_{\mathrm{orb}}$$

where the power of t is keeping track of the codimensions in our orbit decomposition. For example, in the context of $SU(2,2)$ in [9, Table 3], we have

$$s_2 s_1 s_3 s_2 = t^0 \delta_{21} + t^1(\delta_{19} + \delta_{18}) + t^2(\delta_{14} + \delta_{13}) + t^3 \delta_7 + t^4 \delta_1 \in \mathcal{M}_{\mathrm{orb}}.$$

The general strategy will be to parametrize a set $\mathcal{D}_S$ which contains (and sometimes equals) $\mathcal{D}_{\mathrm{hw}} \cap \mathcal{P}_{\mathrm{holo},M}$, using a parameter set $\mathcal{S}$. Now, $\mathcal{S}$ will be in one-to-one correspondence with a subdiagram of $W^{\mathbf{Q}^\sharp}$ and is controlling a set of parameters that simultaneously index $\mathbf{Q}^\sharp$- orbits and $\mathbf{K}$-orbits. Our task, in proving Lemma (4.12), is to understand the $\mathbf{K}$-orbit decomposition of each $\mathcal{O}_w, w \in \mathcal{S}$.

(4.18) Orbits in the HS.1 setting: As discussed in chapter 6, the Kazhdan-Lusztig data for $SU(p', q')$ is implicit in the data for $SU(p,p)$, for any $p \geq p' \geq q'$. For this reason, so far as our paper is concerned, we need only study the quasi-split group $SU(p,p)$. Recall, $n = p + p$. We first parametrize the holomorphic pyramid, via the n-tuples of paired and signed numbers. Given $1 \leq i < j \leq n$, enumerate the numbers in the set $\{k \quad | 1 \leq k \leq n, k \neq i, k \neq j\}$ in increasing order: $i_1 < i_2 < \cdots < i_{n-2}$. Define

$$((i,j)) = (i,j) i_1^+, \ldots, i_{p-1}^+, i_p^-, \ldots, i_{n-2}^-. \tag{4.18a}$$

To each $((i,j))$ we can attach a standard module parameter $\delta_{((i,j))} \in \mathcal{D}_o$ and a $\mathbf{K}$-orbit parameter $\delta_{((i,j)),o} \in V$; for notational simplicity we will usually drop the subscript "o". The totality of all parameters $\delta_{((i,j))}$ as in (4.18a) will parametrize the pyramid $\mathcal{P}_{\mathrm{holo},M}$, together with its $\preceq_w$ order, as in Figure 4.1. Also let

$$((j)) = l_1, l_2, \ldots, l_n, \; 1 \leq j \leq p+1, \tag{4.18b}$$

where $l_k = k^-$ if $k = j$ or $k \geq p+2$ and $l_k = k^+$ otherwise. The parameters $\delta_{((j))}$ index some (but not all) of the discrete series parameters in $\mathcal{D}_o$; equivalently, some of the closed $\mathbf{K}$-orbits in the flag variety. The parameter $\delta_{((p+1))}$ corresponds to a holomorphic discrete series representation. Identifying $\delta_{((i,j))}$ with $((i,j))$, we have a subdiagram of $\hat{\mathcal{P}}_{\mathrm{holo},M}$:

(4.18c)

$$\begin{array}{ccc} & ((p,p+1)) & \\ {}^{s_p}\swarrow & & \searrow^{s_p} \\ ((p)) & & ((p+1)) \end{array}$$

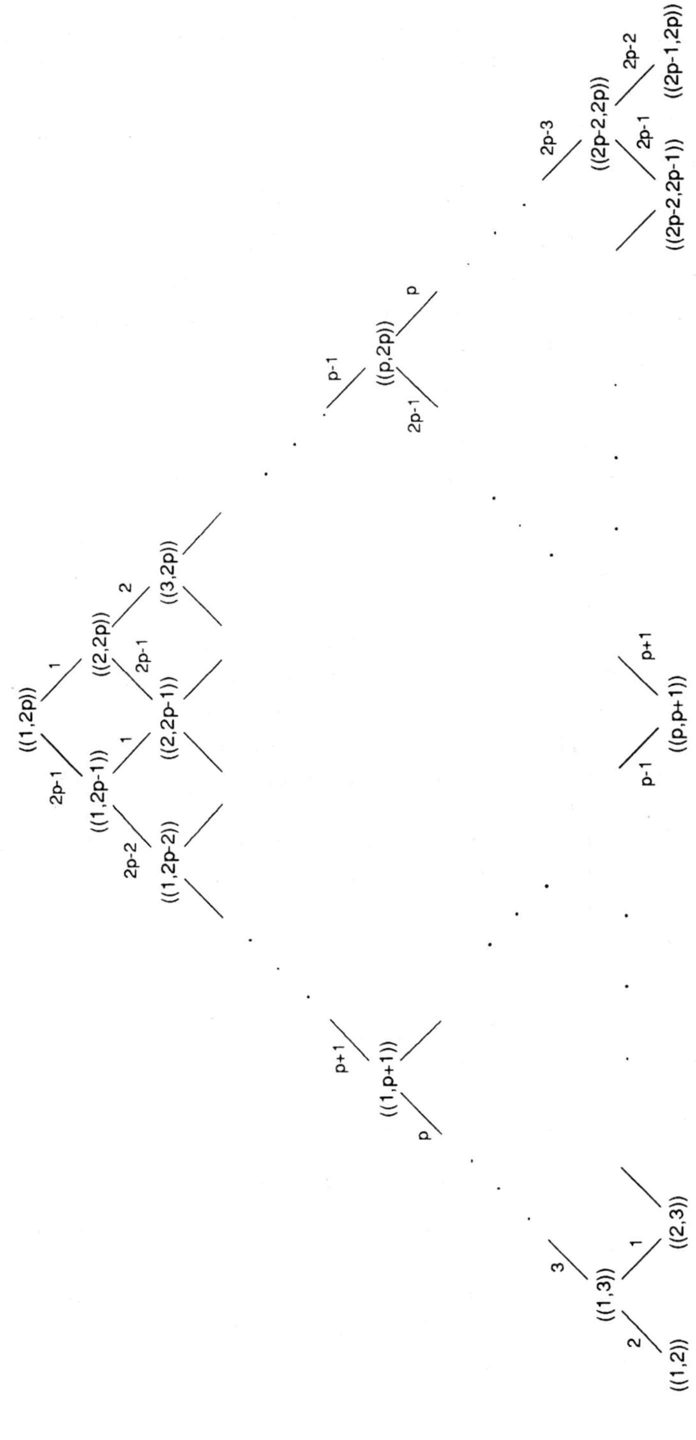

FIGURE 4.1. Holomorphic pyramid in $SU(p,p)$

Recall the $\circ$ and $\times$ actions in chapter 2 (and [23]), then

(4.18d) $$\delta_{((p,p+1))} = s_p \circ \delta_{((p+1))} \quad \text{and} \quad s_p \times \delta_{((p+1))} = \delta_{((p))},$$

since s_p denotes a non-compact Type I reflection.

Consider the set

(4.18e) $$\mathcal{S} = \{((i,j)) \,|\, 1 \le i \le p\,,\, p+1 \le j \le 2p\} \cup \{\, ((p+1))\,\}.$$

Then $\mathcal{S}$ is in one-to-one correspondence with the "lattice $\mathcal{L}$" in [8, (5.8)]; i.e., $\mathcal{S}$ is a subdiagram of the Hermitian symmetric diagram of type HS.1. (In the notation of [8, §5], $\mathcal{S} = \{\, y \le (n,1,1,\ldots,1)\,\}$.) Consequently, to each element of $\mathcal{S}$ we can associate $\mathbf{Q}^\sharp$-orbit parameters $w_{((\ldots))}$ and $\mathbf{K}$-orbit parameters $\delta_{((\ldots))}$. Let $\mathcal{D}_\mathcal{S}$ denote the totality of $\mathbf{Q}^\sharp$-orbit or $\mathbf{K}$-orbit parameters attached to $\mathcal{S}$ in this way. Notice that

(4.18f) $$\mathcal{D}_\mathcal{S} = \mathcal{D}_{\mathsf{hw}} \cap \mathcal{P}_{\mathsf{holo},M}.$$

By (4.17), we need to understand (4.4) in the context of orbits $\mathcal{O}_w$ attached to $w \in \mathcal{D}_\mathcal{S}$. We arrive at the following $\mathcal{M}_{\mathsf{orb}}$ decompositions in (4.4):

$$w_{((p+1))} = \delta_{((p+1))};$$

if $1 \le j \le p$, $0 \le k \le p-1$,

$$\begin{aligned} w_{((j,p+1+k))} = \delta_{((j,p+1+k))} &+ \sum_{1 \le i \le p-j} t^i s_{p+k} \circ \cdots \circ s_{p+1} \circ \delta_{((j,p+1-i))} \\ &+ t^{p+1-j} s_{p+k} \circ \cdots \circ s_{p+1} \circ \delta_{((j))} \\ + \sum_{1 \le l \le k, 1 \le i \le p-j} & t^{i+l} s_{p+k} \circ \ldots s_{p+l} \times s_{p+l-1} \times \ldots s_{p+1} \times \delta_{((j,p+1-i))} \\ + \sum_{1 \le l \le k, 1 \le i \le p-j} & t^{p+1-j+l} s_{p+k} \circ \ldots s_{p+l} \times s_{p+l-1} \times \ldots s_{p+1} \times \delta_{((j))}. \end{aligned}$$

(4.19) Orbits in the HS.2 setting: In this situation, $G = SO_e(2, 2n-1)$ and the parametrization of $\mathcal{D}_o$ is given in (2.8) as follows:

(4.19a) $$\mathcal{D}_o = \mathcal{D}_G \cup \mathcal{D}_{P_n} \cup \mathcal{D}_{P_{n-1}} \cup \mathcal{D}_{P_m},$$

leading to parameters of types i^* in $\mathcal{D}_G$; (i,j) and $\overline{(i,j)}$ in $\mathcal{D}_{P_n}$; $\hat{i}$ in $\mathcal{D}_{P_{n-1}}$; $(i,j)^\sharp$ and $(i,j)^\flat$ in $\mathcal{D}_{P_m}$. The set

(4.19b) $$\mathcal{S} = \{1^*\} \cup \{(0,j) \,|\, 1 \le j \le 2n-2\}$$

is in one-to-one correspondence with the diagram $W^{\mathbf{Q}^\sharp}$ with its longest element removed. Consequently, to each element of $\mathcal{S}$ we can associate $\mathbf{Q}^\sharp$-orbit parameters $w_{(\ldots)}$ and $\mathbf{K}$-orbit parameters $\delta_{(\ldots)}$. Let $\mathcal{D}_\mathcal{S}$ denote the totality of $\mathbf{Q}^\sharp$-orbit or $\mathbf{K}$-orbit parameters attached to $\mathcal{S}$ in this way. Notice that

(4.19c) $$\mathcal{D}_\mathcal{S} = \mathcal{D}_{\mathsf{hw}} \setminus \{\delta_{(1,n-1)^\sharp}\}$$

We can arrange things so that $\delta_{1\bullet}$ indexes the holomorphic discrete series representation in $\mathcal{HC}_o$. The parameters $(\dots)^\sharp$ and $(\dots)^\flat$ are both associated to the same **K**-orbit, as discussed in (2.8). Consequently, $\delta_{(\dots)^\sharp,o} = \delta_{(\dots)^\flat,o}$; we will use the notation $\delta_{(\dots)^\sharp}$ to denote this common **K**-orbit parameter. Recall (4.17) and arrive at the $\mathcal{M}_{\text{orb}}$ decompositions in Proposition (4.4):

$$w_{1\bullet} = \delta_{1\bullet};$$

if $1 \le i \le 2n-2$,

$$w_{(0,i)} = \sum_{0 \le j \le \min(i-1, 2n-i-2)} t^j \delta_{(j,i)} + \sum_{0 \le j \le n-i-2} t^{i+j} \delta_{(i+1,n-i-1-j)^\sharp} + \sum_{i+1 \le j \le 2n-2-i} t^j \delta_{\widetilde{(i,j)}} + t^4 \delta_{\widetilde{(n-i)}} + t^{2n-i-1} \delta_{(2n-i)^\bullet}.$$

(4.20) Orbits in the HS.3 setting: In this case, $G = Sp_n\mathbb{R}$. As remarked in chapter 3, $\mathcal{D}$ decomposes into several blocks. We will only discuss how to describe the holomorphic pyramid, which must lie in $\mathcal{D}_o$ by the arguments used in (2.14). First, let's again emphasize that there are two classes of maximal cuspidal parabolic subgroups, denoted P_1 and P_2 after the statement of Theorem D. By (2.4), $\mathcal{P}_{\text{holo},M_1} = W^{\mathbf{P}_1}_{\text{upper}}$ and $W^{\mathbf{P}_1}_{\text{upper}}$ is particularly simple, arising from a Hermitian pair of type (C_n, C_{n-1}). If we ignore the bottom row in [7, Fig. 4.3], we obtain the parameter set

(4.20a)
$$\mathcal{P}_{\text{holo},M_1} = \{\delta_{(i,1)} \mid 1 \le i \le n\}$$

which appears in Figure 4.2, with the $\preceq_w$ order relations. The more serious issue is the structure of $\hat{\mathcal{P}}_{\text{holo},M_1}$. By (2.4), we see that the pyramids $\mathcal{P}_{\sigma,M_2} = W^{\mathbf{P}_2}_{\text{upper}}$ arise from pairs of type $(C_n, A_1 \times C_{n-2})$. If we replace "$n+1$" by "$n$" and ignore the bottom row of [7, Fig. 4.5], we obtain the poset $\mathcal{P}_{\sigma,M_2}$ and the weak $\preceq$ ordering. In particular, notice that

(4.20b)
$$\mathcal{R}_\sigma = \{(i,j) \mid 2 \le j \le i \le n\}$$

parametrizes a subset of $\mathcal{P}_{\sigma,M_2}$. Let δ_0 index the holomorphic discrete series in $\mathcal{HC}_o$ and define discrete series parameters via successive applications of the cross operation:

(4.20c)
$$s_i \times \cdots \times s_1 \times \delta_0 = \delta_i,\ 1 \le i \le n.$$

Then $\{\delta_i \mid 0 \le i \le n\} \subset \mathcal{D}_{\text{ds}}$. Put

(4.20d)
$$\mathcal{S} = \{0\} \cup \{(i,1) \mid 1 \le i \le n\},$$

then to each element of $\mathcal{S}$ we can associate $\mathbf{Q}^\sharp$-orbit parameters $w_{(\dots)}$ and **K**-orbit parameters $\delta_{(\dots)}$. Let $\mathcal{D}_\mathcal{S}$ denote the totality of $\mathbf{Q}^\sharp$-orbit or **K**-orbit parameters attached to $\mathcal{S}$ in this way. Notice that

(4.20e)
$$\mathcal{D}_\mathcal{S} = \mathcal{P}_{\text{holo},M} \cup \{\delta_0\}.$$

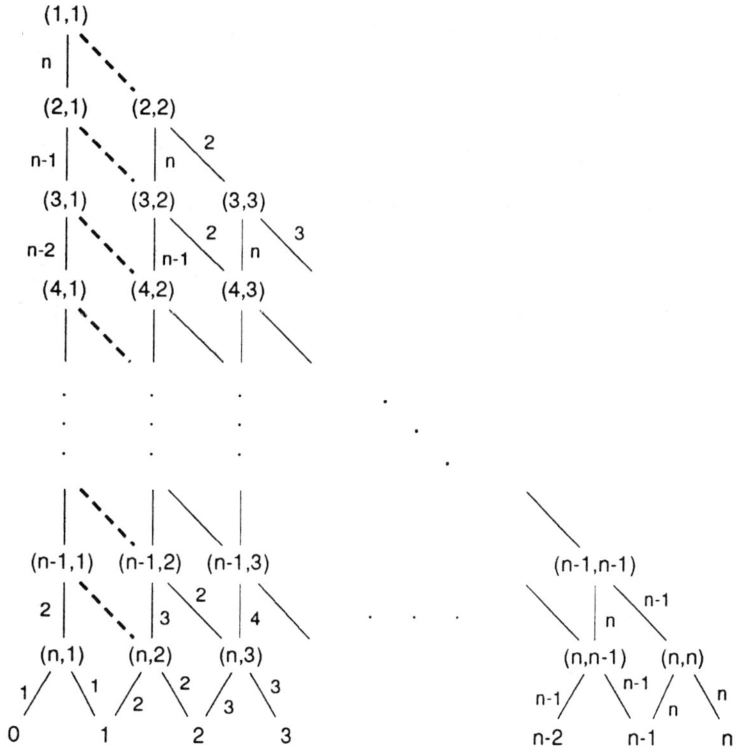

FIGURE 4.2. Holomorphic G-pyramid in $Sp_n\mathbb{R}$

Our assertion is that for the correct choice of $\sigma \in (M_2)_{\mathrm{ds}}$,

$$\hat{\mathcal{P}}_{\mathrm{holo},M_1} = \mathcal{P}_{\mathrm{holo},M_1} \cup \{\delta_{(i,j)} \mid (i,j) \in \mathcal{R}_\sigma\} \cup \{\delta_i \mid 0 \le i \le n\}, \tag{4.20f}$$

which is pictured in Figure 4.2. The justification for this parametrization is roughly as follows: First, obviously, $\mathcal{P}_{\mathrm{holo},M_1} \cup \{\delta_0, \delta_1\} \subset \hat{\mathcal{P}}_{\mathrm{holo},M_1}$. Now, consider the character identity involving the two discrete series parameters δ_1 and $s_2 \times \delta_1$ and $s_2 \circ \delta_1 = \delta_{(n,2)}$. Via the theory in [24], it is an easy matter to check that the Langlands data for $\delta_{(n,2)}$ involves P_2 and some $\sigma \in (M_2)_{\mathrm{ds}}$; take this σ in $\mathcal{R}_\sigma$ above. By the definition of the order on $\mathcal{D}$, we have a G-order relation $\delta_{(n,2)} \preceq \delta_{(n-1,1)}$; hence, $\delta_{(n,2)} \in \hat{\mathcal{P}}_{\mathrm{holo},M_1}$. Continuing in this way, using (4.26c), we see that all the parameters in Figure 4.2 index parameters in $\hat{\mathcal{P}}_{\mathrm{holo},M_1}$. A little more work (involving tau invariant considerations) will insure we pick up no additional parameters.

Recall (4.17) and arrive at the $\mathcal{M}_{\mathrm{orb}}$ decompositions in (4.4):

$$w_0 = \delta_0;$$

if $1 \le j \le n$,

$$w_{(j,1)} = \sum_{0 \le i \le j-1} t^i \delta_{(n+1-j+i,1+i)} + t^j \delta_j$$

(4.21) Orbit decompositions in the HS.4 setting: In this case, $G = SO_e(2, 2n-2)$ and recall the discussion in (2.10) where we described $\mathcal{D}_o$. Using those parameters, define

$$\mathcal{S} = \{\hat{1}\} \cup \{(n-1, j^*) \mid 0 \le j \le n-2\} \cup \{(n-1, j) \mid 0 \le j \le n-2\} \cup \{(1, n-1)^\sharp\}.$$

To each element of $\mathcal{S}$ we can associate $\mathbf{Q}^\sharp$-orbit parameters $w_{(\dots)}$ and $\mathbf{K}$-orbit parameters $\delta_{(\dots)}$. Let $\mathcal{D}_{\mathcal{S}}$ denote the totality of $\mathbf{Q}^\sharp$-orbit or $\mathbf{K}$-orbit parameters attached to $\mathcal{S}$ in this way. Also note that $\delta_{\hat{1}}$ denotes the holomorphic discrete series representation in $\mathcal{HC}_o$. Recalling (4.17), we arrive at the $\mathcal{M}_{\mathrm{orb}}$ decompositions of Proposition (4.4):

$$w_{\hat{1}} = \delta_{\hat{1}};$$

if $2 \le i \le n-1$,

$$w_{(n-1,(n-i)^*)} = \sum_{0 \le j \le i-2} t^j \delta_{(n-j-1,(n-i)^*)} + t^{i-1} \delta_{\hat{i}};$$

$$w_{(n-1,0^*)} = \sum_{0 \le j \le n-2} t^j \delta_{(n-j-1,0^*)} + t^{n-1} \delta_{\widehat{(n+1)}};$$

$$w_{(n-1,0)} = \sum_{0 \le j \le n-2} t^j \delta_{(n-j-1,0)} + t^{n-1} \delta_{\hat{n}};$$

$$\begin{aligned} w_{(n-1,i-1)} = {} & \sum_{0 \le j \le n-1-i} t^j \delta_{(n-j-1,i-1)} + \sum_{0 \le j \le i-2} t^{n-i+j} \delta_{(n+1-i,i-1-j)^\sharp} \\ & + \sum_{0 \le j \le i-2} t^{n-i+1+j} \delta_{\overline{(i-1,i-2-j)}} \\ & + \sum_{0 \le j \le i-2} t^{n-1+i-2-j} \delta_{\overline{(i-1,(i-2-j)^*)}} + t^{n+i-2} \delta_{\widehat{(n+i)}}; \end{aligned}$$

$$\begin{aligned} w_{(1,n-1)^\sharp} = {} & \sum_{0 \le j \le n-2} t^j \delta_{(1,n-1-j)^\sharp} + \sum_{0 \le j \le n-2} t^{j+1} \delta_{\overline{(n-1,n-2-j)}} \\ & + \sum_{0 \le j \le n-2} t^{2n-3-j} \delta_{\overline{(n-1,(n-2-j)^*)}} + t^{2n-2} \delta_{\widehat{2n}}. \end{aligned}$$

(4.22) Orbit decompositions in the HS.5 setting: When $G = SO^*(2n)$, we recall the parameter set for $\mathcal{D}_o$ in (2.11). Define the sets

(4.22a)
$$\begin{aligned} \mathcal{R}_1 &= \{((i,j)) \mid 1 \le i < j \le n\} \cup \{\langle i,j \rangle \mid 1 \le i < j \le n\}, \\ \mathcal{R}_2 &= \{i \mid 1 \le i \le n\}, \\ \mathcal{R} &= \mathcal{R}_1 \cup \mathcal{R}_2. \end{aligned}$$

The set $\mathcal{R}_1$ describes the holomorphic pyramid $\mathcal{P}_{\mathrm{holo},M}$ in Figure 4.3. The parameters in $\mathcal{R}_2$ correspond to certain (but far from all) of the discrete series representations. The discrete series δ_1 will be holomorphic with $s_1 \circ \delta_1 = \delta_{\langle 1,2 \rangle}$.

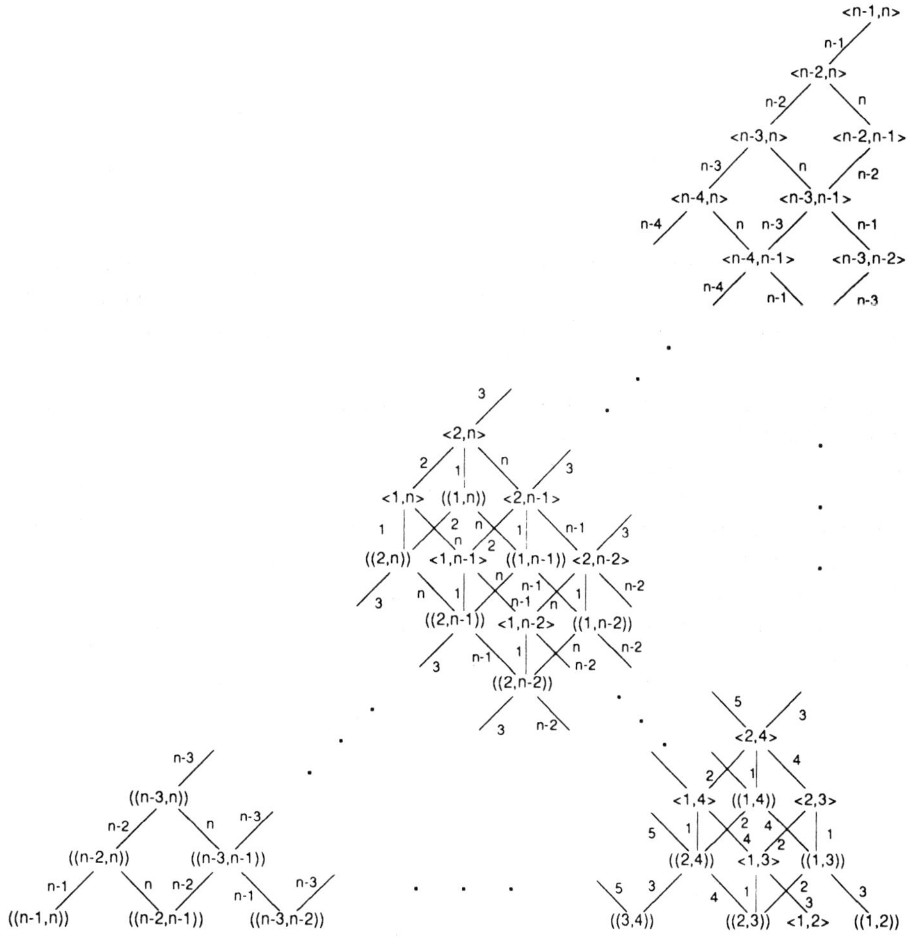

FIGURE 4.3. Holomorphic pyramid for $SO^*(2n)$

Now, consulting Figure 4.3, consider the set $\mathcal{S}'$ consisting of all $z \in \mathcal{P}_{\text{holo},M}$ such that $\langle 1,2\rangle \preceq z \preceq \langle n-1,n\rangle$. Then define

$$\mathcal{S} = \mathcal{S}' \cup \{1\}. \tag{4.22b}$$

Then $\mathcal{S}$ parametrizes a small part of the Hermitian symmetric poset $W^{\mathbf{Q}^\sharp}$ attached to the pair (D_n, A_{n-1}). To each element of $\mathcal{S}$ we can associate $\mathbf{Q}^\sharp$-orbit parameters $w_{(\dots)}$ and $\mathbf{K}$-orbit parameters $\delta_{(\dots)}$. Let $\mathcal{D}_{\mathcal{S}}$ denote the totality of $\mathbf{Q}^\sharp$-orbit or $\mathbf{K}$-orbit parameters attached to $\mathcal{S}$ in this way. Recalling (4.17), we arrive at the $\mathcal{M}_{\text{orb}}$ decompositions of Proposition (4.4):

$$w_1 = \delta_1;$$

if $2 \le j \le n$,

$$w_{\langle 1,j\rangle} = \delta_{\langle 1,j\rangle} + \sum_{1\le i\le j-2} t^i \delta_{((1+i,j))} + t^{j-1}\delta_j;$$

if $3 \le j \le n$,

$$w_{\langle 2,j\rangle} = \delta_{\langle 2,j\rangle} + t^1 s_2 \circ \delta_{((2,j))} + \sum_{2\le i\le j-2} t^i s_2 \circ \delta_{((1+i,j))}$$
$$+ \sum_{2\le i\le j-2} t^{i+1} s_2 \times \delta_{((1+i,j))} + t^{j-1} s_2 \circ \delta_j + t^j s_2 \times \delta_j .$$

This information gives us the orbit decomposition for some of the parameters in $\mathcal{D}_{\mathcal{S}}$. For the others, recalling (4.12), we need only keep track of the $\mathbf{K}$-orbits of codimension 0 and 1 in the remaining $\mathcal{O}_w$. This follows from the above calculations and Figure 4.3, using the algorithm of [9].

(4.23) Orbit decompositions in the HS.6, HS.7 settings: In the two exceptional cases, we explicity compute the Proposition (4.4) decompositions using the parameterizations of $\mathcal{D}_o$ in (2.12).

In the $E_{6,-14}$ case, recall the data in Table 9.1. There is a unique class of maximal cuspidal parabolic subgroup $P = MAN$, which has $M_{ss} = SU(5,1)$. From general theory or [7], there are six classes of discrete series in $M_{\rm ds}$; let's index these as $\{\sigma_1, \dots, \sigma_6\}$. We can and do arrange this numbering so that σ_6 is holomorphic and σ_1 is anti-holomorphic. Let $\mathcal{P}_{\sigma_i,M}$ denote the corresponding pyramids, which are poset isomorphic to the upper halves of [2, Fig. 5.1]. Consequently, $\mathcal{P}_{\rm holo,M} = \mathcal{P}_{\sigma_6,M}$. Recall the unique top parameters $\delta_{\sigma_i,M}$ for these pyramids. Then we can and do arrange things so that using the data of Table 9.1 we have

(4.23a)
$$\begin{aligned}
\delta_{477} &= \delta_{\sigma_6,M};\\
\delta_{475} &= \delta_{\sigma_5,M} = s_6 \times \delta_{\sigma_6,M};\\
\delta_{474} &= \delta_{\sigma_4,M} = s_5 \times s_6 \times \delta_{\sigma_6,M};\\
\delta_{470} &= \delta_{\sigma_3,M} = s_4 \times s_5 \times s_6 \times \delta_{\sigma_6,M};\\
\delta_{460} &= \delta_{\sigma_2,M} = s_3 \times s_4 \times s_5 \times s_6 \times \delta_{\sigma_6,M};\\
\delta_{455} &= \delta_{\sigma_1,M} = s_1 \times s_3 \times s_4 \times s_5 \times s_6 \times \delta_{\sigma_6,M}.
\end{aligned}$$

Recall [10, Fig. 7.1], which parametrizes $W^{\mathbf{Q}^\sharp}$. Taking care to convert the Dynkin diagram labeling of [10, p.81] to our conventions in (2.12) we index the following subset of $W^{\mathbf{Q}^\sharp}$:

$$\mathcal{S} = \{ i^* \mid 0 \le i \le 22 \}.$$

To each element of $\mathcal{S}$ we can associate $\mathbf{Q}^\sharp$-orbit parameters $w_{(\dots)}$ and $\mathbf{K}$-orbit parameters $\delta_{(\dots)}$. Let $\mathcal{D}_{\mathcal{S}}$ denote the totality of $\mathbf{Q}^\sharp$-orbit or $\mathbf{K}$-orbit parameters attached to $\mathcal{S}$ in this way. In particular, we can arrange this so that $\delta_{0^*} = \delta_{27}$ denotes the holomorphic discrete series representation in $\mathcal{HC}_o$. Using Table 9.1, one can check $\mathcal{P}_{\rm holo,M} = \{\delta_i \mid i \in \mathcal{X}_6\}$, where

(4.23b)
$$\begin{aligned}
\mathcal{X}_6 = \{&58,\ 59,\ 60,\ 61,\ 62,\ 63,\ 99,\ 100,\ 101,\ 102,\ 103,\ 144,\\
&145,\ 146,\ 147,\ 148,\ 195,\ 196,\ 197,\ 198,\ 199,\ 248,\ 249,\ 250,\\
&251,\ 298,\ 299,\ 300,\ 346,\ 347,\ 348,\ 390,\ 391,\ 425,\ 453,\ 477\}
\end{aligned}$$

We arrive at the $\mathcal{M}_{\mathrm{orb}}$ decompositions of (4.4); these are given in Lemma (7.7).

For $E_{7,-25}$, we omit the tabulation; it depends on [10, Fig. 7.2], Table 9.2, and the parameters for $\mathcal{D}^{\max}$. The corrresponding set $\mathcal{S}$ will be all parameters below "39" in [10, Fig. 7.2]. The $\preceq_w$ order structure of the holomorphic pyramid is given as the upper half of [2, Fig. 5.2].

5. THE $SO_e(2, N)$ CASES

In this chapter, we verify Theorem A when $G = SO_e(2, N)$.

If $N = 2n - 2$, then as reviewed in (2.15), there is exactly one conjugacy class of maximal cuspidal parabolic subgroups P_n and exactly two pyramids $\mathcal{P}_{+,M_n}$ and $\mathcal{P}_{-,M_n}$ attached to P_n. One of these pyramids will be holomorphic with respect to our complex structure $\mathbf{Q}^\sharp$ and the other is holomorphic for $\bar{\mathbf{Q}}^\sharp$. We may apply Theorem 4.8, which describes a radical filtration for every standard module indexed by $\mathcal{P}_{\text{holo},M_n}$, hence for every module in $\mathcal{F}_{\max}$ by [5]; in particular, multiplicity one holds.

If $N = 2n - 1$, then as reviewed in (2.13), there are two conjugacy classes of maximal cuspidal parabolic subgroups in G, denoted P_n and P_{n-1}. Again, there are exactly two pyramids $\mathcal{P}_{+,M_n}$ and $\mathcal{P}_{-,M_n}$ attached to P_n. One of these pyramids will be holomorphic with respect to our complex structure $\mathbf{Q}^\sharp$ and the other is holomorphic for $\bar{\mathbf{Q}}^\sharp$. For any generalized principal series induced from P_n, we may apply Theorem 4.8 and obtain multiplicity one. In the case of P_{n-1}, $(M_{n-1})_{ss} = SO(1, 2n-2)$, which is neither connected nor Hermitian symmetric. Because of the disconnectedness of the underlying Cartan subgroup, there is just one pyramid attached to P_{n-1}. Now, recalling the notation in (2.13), the parameters involving P_{n-1} in their Langlands data were indexed by $\mathcal{D}_{P_{n-1}} = \{\delta_{\hat{i}} : 1 \leq i \leq n\}$. Thus, the following lemma verifies multiplicity one for the corresponding standard modules and completes the proof of Theorem A in this case. Recall Figure 2.1.

LEMMA 5.1. *Let $G = SO_e(2, 2n-1)$.*
(a) The radical filtration of $\pi(\delta_{\hat{1}})$ is given by a character identity as in (2.10).
(b) If $2 \leq i \leq n$, then the radical filtration of $\pi(\delta_{\hat{i}})$ is given as follows:

$\overline{\pi}(\delta_{\hat{i}})$
$\overline{\pi}(s_i \circ s_{i-1} \circ \cdots \circ s_2 \circ \delta_{n\bullet}) \oplus \overline{\pi}(s_i \circ s_{i-1} \circ \cdots \circ s_2 \circ \delta_{(n+1)\bullet}) \oplus \overline{\pi}(\delta_{\widehat{i-1}})$
$\overline{\pi}(s_{i-1} \circ s_{i-2} \circ \cdots \circ s_2 \circ \delta_{n\bullet}) \oplus \overline{\pi}(s_{i-1} \circ s_{i-2} \circ \cdots \circ s_2 \circ \delta_{(n+1)\bullet})$

Proof. Use the Vogan calculus together with the structure of the diagram $\mathcal{D}$ as described in (2.13) and the inductive observation that the "$\mathcal{D}$-diagram" attached to $SO_e(2, 2n-3)$ will be a subset of $\mathcal{D}$. □

6. THE $SU(p,q)$ CASE

In this chapter we prove Theorem E, by giving explicit formulas for the composition factors of the P-generalized principal series representations in $\mathcal{F}_P$, when P is a maximal cuspidal parabolic subgroup of $SU(p,q)$. As a corollary we obtain Theorem A (that $\mathcal{F}_P$ is multiplicity free) for the case $G = SU(p,q)$. Our result generalizes the "orthogonal sets of roots" formulas of Enright-Shelton for the composition factors of generalized Verma modules associated to a Hermitian symmetric space [11].

Assume that $G = SU(p,q)$ with $p \geq q \geq 1$, and set $n = p+q$. Recall the parametrization of the partially ordered set $\mathcal{D}_o$ from (2.7). Throughout this section, we make the convention that the symbols ϵ, μ, and ν, when used as superscripts, denote elements of $\{+,-\}$.

Recall the subset $\mathcal{D}_o^{\max}$ from (2.4b), and put $\hat{\mathcal{D}}_o^{\max} = \{\gamma \in \mathcal{D}_o \mid \gamma \preceq \delta$ for some $\delta \in \mathcal{D}_o^{\max}\}$. A crucial observation is that the composition factors of $\pi(\delta)$, $\delta \in \mathcal{D}_o^{\max}$, will be of the form $\overline{\pi}(\gamma), \gamma \in \hat{\mathcal{D}}_o^{\max}$. Because of this fact, we require an explicit description of the set $\hat{\mathcal{D}}_o^{\max}$ and the G-order thereon, which will be used constantly in the sequel.

LEMMA 6.1. *(a) $\hat{\mathcal{D}}_o^{\max}$ consists of all parameters in $\mathcal{D}_o$ of the form*

$$(6.2) \qquad \delta = \ \ldots\ (a_1 b_1)\ \ldots\ (a_2 b_2)\ \ldots\ \cdots\ \ldots\ (a_r b_r)\ \ldots$$

(where $\ldots$ denotes a sequence of unpaired numbers) with $r \geq 0$ and $a_1 < b_1 < a_2 < b_2 < \cdots < a_r < b_r$.

(b) Let δ be as in part a), and suppose $i^\epsilon, (i+1)^{-\epsilon} \in \delta$ with $a_k < i < i+1 < b_k$ (some $1 \leq i \leq n-1$, $1 \leq k \leq r$, $\epsilon \in \{+,-\}$). Thus

$$\delta = \cdots (a_1 b_1) \cdots (a_k b_k) \cdots i^\epsilon (i+1)^{-\epsilon} \cdots (a_{k+1}, b_{k+1}) \cdots (a_r b_r) \cdots .$$

Define $\delta \star r_i$ by replacing $(a_k b_k) i^\epsilon (i+1)^{-\epsilon}$ in δ with $(a_k, i)(i+1, b_k)$; i.e.

$$\delta \star r_i = \cdots (a_1 b_1) \cdots (a_k, i) \cdots (i+1, b_k) \cdots (a_{k+1}, b_{k+1}) \cdots (a_r b_r) \cdots .$$

Then $\delta \star r_i \preceq \delta$, and the Bruhat G-order on $\hat{\mathcal{D}}_o^{\max}$ is the smallest order containing all such relations as well as the weak order.

Proof. Let $\mathcal{A}$ denote the subset of $\mathcal{D}_o$ consisting of all parameters δ of the form (6.2). We claim that if $\delta \in \mathcal{A}, \gamma \in \mathcal{D}_o$, $l'(\gamma) = l'(\delta) - 1$, and $\gamma \notin s \circ \delta$ for any $s \in S$, then $\gamma \preceq \delta$ if and only if $\gamma = \delta \star r_i$ for some i as in part (b). Assume the claim for the moment. In particular, it is clear (using (2.7)) that if $l^I(\gamma) = l^I(\delta) - 1$, and $\gamma \preceq \delta$, then $\gamma \in \mathcal{A}$. Now it follows from the work of Richardson-Springer [18] that the G-order on $\mathcal{D}_o$ for $SU(p,q)$ satisfies what they call the "chain condition." That is, the order is generated by the relations $\gamma \preceq \delta$, $l^I(\gamma) = l^I(\delta) - 1$. This will complete the proof of (b) (for $\delta \in \mathcal{A}$). For (a), observe that $\mathcal{D}_o^{\max}$ consists of those parameters $\delta \in \mathcal{A}$ for which $r = 1$ in (6.2). It now follows from part (b) that $\hat{\mathcal{D}}_o^{\max} \subseteq \mathcal{A}$. Finally it is easy to see by induction on r that $\mathcal{A} \subseteq \hat{\mathcal{D}}_o^{\max}$.

It remains only to prove the claim above. We do this by induction on $l^I(\delta)$. Let $\delta \in \mathcal{A}$, $\gamma \in \mathcal{D}_o$, with $l^I(\gamma) = l^I(\delta) - 1, \gamma \notin s \circ \delta$ for any $s \in S$, and $\gamma \preceq \delta$. Then there exists $s \in S$, $\gamma', \delta' \in \mathcal{D}_o$ such that $\delta \xrightarrow{s} \delta', \gamma \xrightarrow{s} \gamma'$, and $\gamma' \preceq \delta'$. Examining the possibilities for $\delta \xrightarrow{s} \delta'$ one finds that $\delta' \in \mathcal{A}$. Since $l^I(\delta') = l^I(\delta) - 1$ and $l^I(\gamma') = l^I(\delta') - 1$, the induction hypothesis implies that either $\gamma' \in t \circ \delta'$ for some $t \in S$, or $\gamma' = \delta' \star r_i$ for some i. Suppose the first case. If s and t are reflections through orthogonal roots, then $\gamma \in t \circ \delta$, contrary to assumption. So assume the roots which determine s and t are non-orthogonal. We have the following possibilities.

(6.3a) $$\delta = \cdots (i-1, b)\ i^{\epsilon}\ (i+1)^{-\epsilon} \cdots, \quad b > i+1$$

$$s = s_{i-1} \swarrow$$

$$\delta' = \cdots (i-1)^{\epsilon} (i, b) (i+1)^{-\epsilon} \cdots \qquad \gamma = \cdots (i-1, i)(i+1, b) \cdots$$

$$t = s_i \searrow \qquad \swarrow s = s_{i-1}$$

$$\gamma' = \cdots (i-1)^{\epsilon}\ i^{-\epsilon}\ (i+1, b) \cdots$$

(so $\gamma = \delta \star r_i$)

(6.3b) $$\delta = \cdots (a, i+2) \cdots i^{\epsilon}\ (i+1)^{-\epsilon} \cdots, \quad a < i$$

$$s = s_{i+1} \swarrow$$

$$\delta' = \cdots (a, i+1) \cdots i^{\epsilon} (i+2)^{-\epsilon} \cdots \qquad \gamma = \cdots (a, i) \cdots (i+1, i+2) \cdots$$

$$t = s_i \searrow \qquad \swarrow s = s_{i+1}$$

$$\gamma' = \cdots (a, i) \cdots (i+1)^{\epsilon} (i+2)^{-\epsilon} \cdots$$

(so again $\gamma = \delta \star r_i$)

(6.3c)

$$\begin{array}{ccc} & \delta = \cdots(i-1,i+1)i^{\epsilon}\cdots & \\ {}^{s=s_{i-1}}\swarrow & & \searrow^{t=s_i} \\ \delta' = \cdots(i-1)^{\epsilon}(i,i+1)\cdots & & \gamma = \cdots(i-1,i)(i+1)^{\epsilon} \\ {}_{t=s_i}\searrow & & \swarrow_{s=s_{i-1}} \\ & \gamma' = \cdots(i-1)^{\epsilon}i^{-\epsilon}(i+1)^{\epsilon}\cdots & \end{array}$$

(so $\gamma = s_i \circ \delta$)
and similarly reversing the roles of γ and δ', s and t in (6.3c). Notice that we have proved the existence of the relation $\delta \star r_i \preceq \delta$ when either $i = a_k + 1$ (in (6.3a)) or $i = b_k - 2$ (in (6.3b)).

Now suppose we are in the second case: $\gamma' = \delta' \star r_i$. It is evident that s cannot affect coordinates i or $i+1$ in δ, and hence that $\gamma = \delta \star r_i$. To see that $\delta \star r_i \preceq \delta$ for all i as in the statement of part (b), use induction on $i - a_k$ and the following diagram.

(6.3d)

$$\begin{array}{ccc} & \delta = \cdots(a_k,b_k)(a_k+1)^{\mu}\cdots\ i^{\epsilon}\ (i+1)^{-\epsilon}\cdots, \qquad \mu \in \{+,-\} & \\ {}^{s=s_{a_k}}\swarrow & & \\ \delta' = \cdots a_k^{\mu}(a_k+1,b_k)\cdots i^{\epsilon}(i+1)^{-\epsilon}\cdots & & \gamma = \cdots(a_k,i)(a_k+1)^{\mu}\cdots(i+1,b_k)\cdots \\ {}_{r_i}\searrow & & \swarrow_{s=s_{a_k}} \\ & \gamma' = \cdots a_k^{\mu}(a_k+1,i)\cdots(i+1,b_k)\cdots & \end{array}$$

(where $\gamma' = \delta' \star r_i \preceq \delta'$ by the induction hypothesis).

This completes the proof of the lemma. □

The next ingredient we need is an analog of Enright-Shelton's construction $\overline{wr_\Omega}$ = the projection on $W^{Q^\#}$ of $w\left(\prod_{\beta\in\Omega^+} s_\beta\right)$ for $w \in W^{Q^\#}$ and Ω a set of pairwise orthogonal roots (cf. [11]). The main difficulty to be overcome in the present situation is that the roots in Ω will not, in general, be orthogonal, so the natural analog of $\prod_{\beta\in\Omega} s_\beta$ will depend on the order of the factors.

We begin with a (*set-valued*) "action" $\star$ of the reflections through positive roots on $\hat{\mathcal{D}}_o^{\max}$, which will play the role of s_β. Fix $\delta \in \hat{\mathcal{D}}_o^{\max}$ as in (6.2). Suppose first that $\alpha = \alpha_i$ is a simple root. If α is complex of downward type or real for δ, define $\delta \star r_\alpha = s_\alpha \circ \delta$. If α is noncompact imaginary for δ and i satisfies the conditions of part (b) of Lemma 6.1, define $\delta \star r_\alpha = \delta \star r_i$. (Notice that, in this case, $\delta \star r_\alpha \preceq \delta$ whereas $s_\alpha \circ \delta \succeq \delta$; in fact, $s_\alpha \circ \delta \notin \hat{\mathcal{D}}_o^{\max}$.) For all other simple α, define $\delta \star r_\alpha = \varnothing$.

Now suppose $\beta = e_i - e_j \in \Phi^+$ is *non*-simple (so that $i < j - 1$). The notions of *real*, *imaginary*, *complex*, *compact*, *noncompact*, *upward*, and *downward* can be extended to β in the obvious way: the definitions are exactly the same as in (2.7), with "$i+1$" replaced by "j." Then $\delta \star r_\beta$ is defined as follows.

(i) If β is complex of downward type for δ, interchange i and j in δ.

(ii) If β is real for δ, then

a) if $\delta = \cdots (i,j)(i+1)^\epsilon \cdots (j-1)^{-\epsilon} \cdots$,

(6.4) then $\delta \star r_\beta = \cdots i^\epsilon (i+1)^\epsilon \cdots (j-1)^{-\epsilon} j^{-\epsilon} \cdots$;

b) if $\delta = \cdots (i,j)(i+1)^\epsilon \cdots (j-1)^\epsilon \cdots$, then $\delta \star r_\beta = \varnothing$.

(iii) If $\delta = \cdots (a_k, b_k) \cdots i^\epsilon \cdots j^{-\epsilon} \cdots (a_{k+1}, b_{k+1}) \cdots$, $(a_k < i < j < b_k)$, then $\delta \star r_\beta = \cdots (a_k, i) \cdots (j, b_k) \cdots (a_{k+1}, b_{k+1}) \cdots$.

(iv) Otherwise, $\delta \star r_\beta = \varnothing$.

(Notice that $\delta \star r_\beta$ is at most single-valued if β is non-simple, whereas $\delta \star r_\beta$ is double-valued if β is simple and real for δ.)

Because the $\star$ action is set-valued, we must define an action on subsets of $\mathcal{D}_o^{\max}$. If $\beta \in \Phi^+$ is arbitrary (possibly simple), and $\Gamma \subset \hat{\mathcal{D}}_o^{\max}$, put

$$\Gamma \star r_\beta = \bigcup_{\gamma \in \Gamma} \gamma \star r_\beta. \tag{6.5}$$

Finally, we come to the analog of Enright-Shelton's $\overline{wr_\Omega}$. If $\Omega \subset \Phi^+$ is a non-empty set of positive roots, say $\Omega = \{\beta_1, \beta_2, \cdots, \beta_m\}$, and $\delta \in \hat{\mathcal{D}}_o^{\max}$, we define

$$\delta \star r_\Omega = \bigcap_{\sigma \in S_m} (\cdots ((\delta \star r_{\beta_{\sigma(1)}}) \star r_{\beta_{\sigma(2)}}) \cdots) \star r_{\beta_{\sigma(m)}} \tag{6.6}$$

where S_m is the permutation group on m letters. If $\Omega = \varnothing$, define $\delta \star r_\Omega = \delta$.

We can now describe the collection $\mathcal{E}_\delta$ of sets Ω of roots attached to δ, according to the philosophy outlined in (1.3)–(1.5). Given a subset $\Omega \subseteq \Phi^+$, define $\Omega^+ \subseteq \Omega$ to be the set of maximal elements in Ω, in the usual partial ordering $\leq$ of $\mathfrak{h}^*$. Put $\Omega^- = \Omega \setminus \Omega^+$. For $\gamma \in \Omega$, set $\Omega_\gamma = \{\zeta \in \Omega \mid \zeta \leq \gamma, \zeta \neq \gamma\}$.

Fix $\delta \in \hat{\mathcal{D}}_o^{\max}$, parametrized as in (6.2). For $1 \leq i \leq n$, if $i^\epsilon \in \delta$, set $\mathrm{sgn}_\delta(i) = \epsilon$. Also for $\epsilon \in \{+,-\}$, define $-\epsilon$ so that $\{\epsilon, -\epsilon\} = \{+,-\}$. Suppose $\beta = e_i - e_j \in \Phi^+$ is either complex or noncompact for δ. Define the *parity* of β for δ,

$$\mathrm{par}_\delta(\beta) = \begin{cases} \mathrm{sgn}_\delta(i) & \text{if } i \text{ is unpaired in } \delta \\ -\mathrm{sgn}_\delta(j) & \text{if } j \text{ is unpaired in } \delta. \end{cases} \tag{6.7}$$

Recalling (2.7), we note that this is well-defined because of our assumption on β.

DEFINITION 6.8. Given $\delta \in \hat{\mathcal{D}}_o^{\max}$, let $\mathcal{E}_\delta$ be the collection of all subsets $\Omega \subset \Phi^+$ satisfying the following properties:

(i) Ω contains no compact roots for δ.
(ii) If $\beta \in \Omega^+$ then there exists a real root α_0 for δ (not necesarily in Ω) such that $\beta \leq \alpha_0$. If α_0 is real for δ then $\sum_{\substack{\beta \in \Omega^+ \\ \beta \leq \alpha_0}} \beta \leq \alpha_0$.
(iii) If $\zeta \in \Omega^-$ then there exists $\beta \in \Omega^+$ with $\zeta \in \Omega_\beta$.
(iv) Given $\beta \in \Omega$ let $\zeta, \zeta' \in \Omega_\beta$. Then
  (a) $(\beta, \zeta) = (\beta, \zeta') = 0$
  (b) $(\zeta, \zeta') = 0$
  (c) $\mathrm{par}_\delta(\zeta) = \mathrm{par}_\delta(\zeta')$
  (d) if β is complex or noncompact for δ, then $\mathrm{par}_\delta(\zeta) = -\mathrm{par}_\delta(\beta)$.
(v) Given $\beta \in \Omega$ and $\xi \in \Phi^+$ with $\xi \leq \beta$, $\xi \neq \beta$, and $(\beta, \xi) \neq 0$, there exists $\zeta \in \Omega_\beta$ such that $(\zeta, \xi) \neq 0$.

Shortly we shall examine a number of consequences of this definition, both to better orient the reader and for use in the sequel. For now we shall be satisfied with these observations: (6.8)(iii) is equivalent to [11, (8.2)(iii)] and (6.8)(v) is equivalent to [11, (8.2)(ii)]. The conditions in (6.8)(iv) should be viewed as a partial substitute for orthogonality.

We now have developed all the concepts needed to state the main theorem of this section!

THEOREM E. *Let $G = SU(p,q)$ and assume $\delta \in \hat{\mathcal{D}}_0^{\max}$.*

(1) *$\overline{\pi}(\gamma)$ is a composition factor of $\pi(\delta)$ if and only if $\gamma \in \delta \star r_{\Omega^+}$ for some $\Omega \in \mathcal{E}_\delta$.*
(2) *All the composition factors of $\pi(\delta)$ occur with multiplicity one.*
(3) *The composition factor(s) $\overline{\pi}(\gamma)$ attached to $\Omega \in \mathcal{E}_\delta$ lie(s) in layer $|\Omega^+|$ of the radical filtration of $\pi(\delta)$.*
(4) *$\overline{\pi}(\gamma)$ is a standard factor precisely when $\Omega = \Omega^+$.*

Before giving the proof we shall prove several preliminary results. We begin with the promised consequences of Definition 6.8. Recall that for a positive root $\beta = \sum_{i=1}^{n-1} c_i\alpha_i$, the *support* of β, $\mathrm{supp}\,\beta = \{\alpha_i \mid c_i \neq 0\}$.

PROPOSITION 6.9. *Let $\delta \in \hat{\mathcal{D}}_0^{\max}$ and $\Omega \in \mathcal{E}_\delta$.*

(1) *If $\beta_1, \beta_2 \in \Omega^+$ then $\mathrm{supp}\,\beta_1 \cap \mathrm{supp}\,\beta_2 = \varnothing$. (However, β_1 and β_2 need not be orthogonal.) In particular, for any $\beta \in \Omega^+$, $\Omega_\beta \subset \Omega^-$.*
(2) *Ω^- consists of pairwise orthogonal roots.*
(3) *If $\zeta_1, \zeta_2 \in \Omega^-$ with $\mathrm{supp}\,\zeta_1 \cap \mathrm{supp}\,\zeta_2 \neq \varnothing$, then either $\zeta_1 < \zeta_2$ or $\zeta_2 < \zeta_1$.*
(4) *If $\beta = e_i - e_j \in \Omega$ then $j - i$ is odd.*
(5) *If $\beta = e_i - e_j \in \Omega$ with $i < j-1$, and if $(i+1)^\epsilon, (j-1)^\mu \in \delta$, then $\mu = -\epsilon$.*

Proof. (1) This follows from (6.8)(ii) and (6.1). For if $\delta = \cdots(a_1b_1)\cdots \cdots(a_mb_m)\cdots$ with $a_1 < b_1 < \cdots < a_m < b_m$, then the real roots for δ are the $e_{a_j} - e_{b_j}$, $1 \leq j \leq m$. If β_1 and β_2 are dominated by different real roots for δ, then clearly their supports are disjoint. On the other hand, the summation condition in (6.8)(ii) is equivalent to the statement that the roots in Ω^+ dominated by a fixed real root α_0 have pairwise disjoint supports. (Notice that it is possible to have $\beta_1 = e_i - e_j$ and $\beta_2 = e_j - e_k$, $i < j < k$, as long as there is a real root $e_a - e_b$ for δ with $a \le i < k \le b$; in such a case, $(\beta_1, \beta_2) \ne 0$.)

(2) Let $\zeta_1, \zeta_2 \in \Omega^-$. By (6.8)(iii) there exist $\beta_1, \beta_2 \in \Omega^+$ such that $\zeta_i \in \Omega_{\beta_i}$ $(i = 1, 2)$. If $\beta_1 \ne \beta_2$ then (6.8)(iv)(a) and (1) above imply $(\zeta_1, \zeta_2) = 0$. If $\beta_1 = \beta_2$ we can invoke (6.8)(iv)(b).

(3) Suppose, on the contrary, that $\zeta_1 = e_i - e_k$, $\zeta_2 = e_j - e_l$, $i < j < k < l$. It is clear from (1) and (6.8)(iii) that there is a unique $\beta_1 \in \Omega^+$ with $\zeta_1, \zeta_2 \in \Omega_{\beta_1}$. Now apply (6.8)(v) with $\beta = \zeta_1$, $\xi = e_j - e_k$ to conclude that $\zeta = \pm(\epsilon_j - \epsilon_r) \in \Omega_\beta$ for some $i \le r \le k$. Then $\zeta \in \Omega_{\beta_1}$, but $(\zeta_1, \zeta_2) \ne 0$, contradicting (6.8)(iv)(b).

(4) Effectively (6.8)(v) says that if $\beta = \epsilon_i - \epsilon_j \in \Omega$, then for every k, $i < k < j$, there exists $\pm(e_k - \epsilon_l) \in \Omega_\beta$ (hence $i < l < j$ also). From (6.8)(iii) and (1), $\Omega_\beta \subset \Omega^-$, so by (2), these elements of Ω_β are pairwise orthogonal. Hence there must be an even number of such k, i.e. $j - i$ is odd.

(5) By the remarks in the proof of (4), there exist $e_{i+1} - e_k, e_l - e_{j-1} \in \Omega_\beta$. By (6.8)(iv)(c) and (6.7),

$$\epsilon = \mathrm{par}_\delta(e_{i+1} - e_k) = \mathrm{par}_\delta(e_l - e_{j-1}) = -\mu. \qquad \square$$

The next several results provide specific information about the correspondence between sets $\Omega \in \mathcal{E}_\delta$ and subsets $\delta \star r_{\Omega^+}$. These results will be crucial for proving that the composition factors of $\pi(\delta)$ occur without multiplicities. We begin by showing that Ω^+ uniquely determines Ω.

PROPOSITION 6.10. *Given $\delta \in \hat{\mathcal{D}}_o^{\max}$ and a subset $\Theta \subseteq \Phi^+$, there exists at most one $\Omega \in \mathcal{E}_\delta$ such that $\Omega^+ = \Theta$.*

Proof. Suppose such an Ω exists. We must show that Ω^- is uniquely determined by Ω^+. Fix $\beta = e_i - e_j \in \Theta = \Omega^+$. As in the proof of (6.9)(4), for every k, $i < k < j$, there exists $\zeta = \pm(e_k - e_l) \in \Omega^-$, for some $i < l < j$; by (6.8)(iii) and (6.9)(1), every element of Ω^- occurs in this way. If β is simple then no $\zeta \in \Omega^-$ is dominated by β, so we may assume β is not simple. Then $(i+1)^\epsilon, (j-1)^{-\epsilon} \in \delta$ by (6.9)(5). Furthermore every $\zeta \in \Omega_\beta$ must have parity ϵ. Thus, keeping in mind (6.9)(3), the roots $e_k - e_l \in \Omega_\beta$ $(i < k < l < j)$ must be determined by the following algorithm.

(1) For any k such that $k^\epsilon, (k+1)^{-\epsilon} \in \delta$, $e_k - e_{k+1} \in \Omega_\beta$.
(2) Remove from the sequence δ all numbers k', l' such that $e_{k'} - e_{l'}$ has already been determined to belong to Ω_β. If $k^\epsilon, l^{-\epsilon}$ are adjacent in this shortened sequence, then $e_k - e_l \in \Omega_\beta$.
(3) Repeat step 2 until all numbers between i and j have been paired.

Thus Ω_β, and hence Ω^-, is uniquely determined by Ω^+ and conditions (6.8). $\square$

The next proposition will imply that for "most" parameters $\delta \in \mathcal{D}_o^{\max}$, each $\Omega \in \mathcal{E}_\delta$ corresponds to exactly one composition factor of $\pi(\delta)$.

PROPOSITION 6.11. *(1) Let $\delta \in \mathcal{D}_o^{\max}$, so that $\delta = \cdots(a\ b)\cdots$ contains exactly one pair. Assume that $b > a+1$. Then for every $\Omega \in \mathcal{E}_\delta$, $\delta \star r_{\Omega^+}$ consists of exactly one element of $\mathcal{D}_o$.*

(2) Let δ be as above with $b = a + 1$. If $\Omega \in \mathcal{E}_\delta$, then either $\Omega = \{e_a - e_{a+1}\}$, in which case $\delta \star r_{\Omega^+}$ consists of exactly two elements of $\mathcal{D}_o$, or $\Omega = \varnothing$, in which case $\delta \star r_\Omega = \delta$.

(3) More generally, let $\delta \in \hat{\mathcal{D}}_o^{\max}$ and $\Omega \in \mathcal{E}_\delta$. Then

$$\log_2 |\delta \star r_{\Omega^+}| = \#\{\, 1 \leq i \leq n-1 \mid \alpha_i \in \Omega \text{ and } \alpha_i \text{ is real for } \delta \,\}$$

(where, as usual, α_i denotes a simple root).

We require a lemma.

LEMMA 6.12 (HEREDITARITY). *Let $\delta \in \mathcal{D}_o^{\max}$, $\varnothing \neq \Omega \in \mathcal{E}_\delta$, and $\beta \in \Omega^+$. Let $\delta' \in \delta \star r_\beta$ and put $\Omega' = \Omega \setminus (\Omega_\beta \cup \{\beta\})$. Then $\Omega' \in \mathcal{E}_{\delta'}$.*

Proof. Write $\beta = e_i - e_j$. By (6.8)(ii) there exists a real root $\alpha_0 = e_a - e_b$ for δ, with $a \leq i < j \leq b$. There are four cases.

Case 1: $a < i < j < b$. Then $\delta = \cdots (a\ b) \cdots i^\epsilon \cdots j^{-\epsilon} \cdots$, $\delta' = \cdots (a\ i) \cdots \cdots (j\ b) \cdots$; there may be other pairs in δ but these will be "disjoint" from $(a\ b)$ and will remain unaffected in δ'. It is obvious that Ω' contains no compact roots for δ'. Note that $\Omega'^+ = \Omega^+ \setminus \{\beta\}$. Any elements of Ω'^+ which were dominated by α_0 for δ will be dominated by $e_a - e_i$ or $e_j - e_b$ for δ', by (6.9)(1). For the same reason, the second condition in (6.8)(ii) for Ω and α_0 implies the corresponding condition for Ω' and $e_a - e_k$ and $e_j - e_b$. Any other real root for δ remains real for δ', so (6.8)(ii) is satisfied for Ω'. Next, $\Omega' = \Omega \setminus \Omega_\beta$. Hence (6.8)(iii) is clear for Ω'. To prove (6.8)(iv), note that for $\beta' \in \Omega'$, $\Omega'_{\beta'} = \Omega_{\beta'}$, which makes (a), (b), and (c) obvious. The only roots $\beta' \in \Omega'$ which have different types for δ and δ' change from complex to real ($\beta' = e_a - e_i$ or $e_j - e_b$), or from noncompact to complex ($\beta' = e_k - e_i$, $a < k < i$, or $e_j - e_l$, $j < l < b$). In the former case, $\mathrm{par}_{\delta'}(\beta')$ is undefined, while in the latter case, $\mathrm{par}_{\delta'}(\beta') = \mathrm{par}_\delta(\beta')$. Thus (d) is satisfied. Finally, the fact that $\Omega'_{\beta'} = \Omega_{\beta'}$ makes (6.8)(v) evident for Ω'.

Case 2: $a < i < j = b$. Then $\delta = \cdots (a\ b) \cdots i^\epsilon \cdots$, $\delta' = \cdots (a\ i) \cdots b^\epsilon \cdots$. The analysis proceeds as in Case 1, except that now there are no roots in Ω' dominated by $e_j - e_b$.

Case 3: $a = i < j < b$. Apply Case 2 and symmetry.

Case 4: $i = a$, $j = b$. Then $\delta = \cdots (a\ b) \cdots$, $\delta' = \cdots a^\epsilon \cdots b^{-\epsilon} \cdots$ for some $\epsilon \in \{+,-\}$. All roots in Ω' must be dominated by real roots for δ (and for δ') other than α_0. Hence properties (6.8) for Ω' follow immediately from the corresponding properties for Ω. □

Proof of 6.11. (1) Let $\delta = \cdots (a\ b) \cdots$ with $a < b-1$. (Recall that, in contrast to the situation in the lemma, we are now assuming δ contains a unique pair.) We first show uniqueness, and then existence of an element in $\delta \star r_{\Omega^+}$.

Uniqueness: Suppose that $|\delta \star r_{\Omega^+}| > 1$. Since the $\star$ action is multi-valued only for real simple roots, we can find a subset $\Gamma \subset \Omega^+$ and $\alpha_i \in \Omega^+ \setminus \Gamma$ such that $\delta \star r_\Gamma$ consists of a single element $\gamma = \cdots (i,\ i+1) \cdots$, while $|\delta \star r_{\Gamma \cup \{\alpha_i\}}| = 2$.

Since $a < b - 1$ we may assume without loss of generality that there exists $\beta = e_{i+1} - e_j \in \Gamma$ such that, if we put $\Gamma' = \Gamma \setminus \{\beta\}$ and $\gamma' = \delta \star r_{\Gamma'}$, then either

$$(a)\ \ \gamma' = \cdots (i,j)(i+1)^{\epsilon} \cdots, \quad \text{or}$$
$$(b)\ \ \gamma' = \cdots (i,k)(i+1)^{\epsilon} \cdots j^{-\epsilon} \cdots, \quad j < k.$$

(The symmetrical situation $\beta = e_j - e_i$ follows similarly.)

In (a),

$$\begin{aligned}(\gamma' \star r_\beta) \star r_{\alpha_i} &= \{\cdots (i, i+1) \cdots j^{\epsilon} \cdots\} \star r_{\alpha_i} \\ &= \{\cdots i^{\epsilon}(i+1)^{-\epsilon} \cdots j^{\epsilon} \cdots, \quad \cdots j^{-\epsilon}(i+1)^{\epsilon} \cdots j^{\epsilon} \cdots\}\end{aligned}$$

while

$$(\gamma' \star r_{\alpha_i}) \star r_\beta = \{\cdots i^{\epsilon}(i+1, j) \cdots\} \star r_\beta.$$

If β is simple, the last set is equal to

$$\{\cdots i^{\epsilon}(i+1)^{-\epsilon} \cdots j^{\epsilon} \cdots, \quad \cdots i^{\epsilon}(i+1)^{\epsilon} \cdots j^{-\epsilon} \cdots\}.$$

If β is not simple, then Lemma (6.12) and (6.8)(iv)(d) imply that

$$\gamma' = \cdots (i,j)(i+1)^{\epsilon}(i+2)^{-\epsilon} \cdots (j-1)^{\epsilon} \cdots,$$

hence

$$(\gamma' \star r_{\alpha_i}) \star r_\beta = \{\cdots i^{\epsilon}(i+1)^{-\epsilon} \cdots j^{\epsilon} \cdots\}$$

by (6.4)(ii)(a). In either case, (6.6) implies that $\delta \star r_{\Gamma \cup \{\alpha_i\}}$ consists of (at most) the single element $\cdots i^{\epsilon}(i+1)^{-\epsilon} \cdots j^{\epsilon} \cdots$.

In case (b) a similar (but easier) analysis shows that $\delta \star r_{\Gamma \cup \{\alpha_i\}}$ consists of (at most) the single element $\cdots i^{\epsilon}(i+1)^{-\epsilon} \cdots (j,k) \cdots$. This contradiction proves that $|\delta \star r_{\Omega^+}| \leq 1$.

Existence: We exhibit an element $\gamma \in \hat{\mathcal{D}}_o^{\max}$ such that $\gamma \in (\cdots((\delta \star r_{\beta_1}) \star r_{\beta_2}) \cdots) \star r_{\beta_m}$ for any ordering $\{\beta_1, \beta_2, \ldots, \beta_m\}$ of Ω^+. The following new notation for elements of $\hat{\mathcal{D}}_o^{\max}$ will be more convenient: list the numbers $1, \ldots, n$ in increasing order; denote the sign attached to any single number by a superscript (as usual); and join any paired numbers by a curved line below the numbers. We also represent the elements $e_i - e_j \in \Omega^+$ by "square brackets" joining i and j above the numbers. In this scheme, a typical Ω^+ for δ is denoted:

(6.13)

$$\delta = 1 \cdots a \cdots i_1^{\epsilon} \cdots i_2^{-\epsilon} \cdots i_r^{\pm\epsilon} \cdots j_1^{\mu} \cdots j_2^{-\mu} \cdots j_3^{\mu} \cdots j_s^{\pm\mu} \cdots\cdots z_1^{\pm\nu} \cdots z_{t-1}^{-\nu} \cdots z_t^{\nu} \cdots b \cdots n$$

(We have illustrated the case where Ω^+ contains a root of the form $e_a - e_{i_1}$ but not one of the form $e_{z_t} - e_b$; the other cases are similar.) Using (6.4) and (6.8), one sees that an element γ belonging to $\delta \star r_{\Omega^+}$ is:

(6.14)

$$\gamma = 1 \cdots a^{\epsilon} \cdots i_1^{-\epsilon} \cdots i_2^{\epsilon} \cdots i_r \cdots j_1 \cdots j_2^{\mu} \cdots j_3^{-\mu} \cdots j_s \cdots\cdots\cdots z_1 \cdots z_{t-1}^{\nu} \cdots z_t \cdots b \cdots n$$

(The only tricky point is this: suppose $i_1 > a+1$, and $\beta' = e_{i_1} - e_{i_2}$ occurs in the chosen ordering of Ω^+ prior to $\beta = e_a - e_{i_1}$. Then r_β is applied to a parameter δ' for which β is real; how do we know that δ' is in case (6.4)(ii)(a) and not (6.4)(ii)(b)? Well, since β is complex for δ, conditions (iv)(d) and (v) of (6.8) imply that $(a+1)^\epsilon$, $(i_1-1)^{-\epsilon}$ occur in δ, hence also in δ' (recall the proof of (6.9)(5)). This is precisely what is needed.)

(2) This is clear from (6.8)(ii) and (2.7d4).

(3) Write

$$\Omega = \bigcup_{\alpha\in\Phi^+ \text{ real for } \delta} \Omega(\alpha),$$

where $\Omega(\alpha) = \{\beta \in \Omega \mid \beta \leq \alpha\}$. Then $\Omega(\alpha) \in \mathcal{E}_\delta$, and we can compute $\delta \star r_{\Omega^+}$ by applying the various $r_{\Omega(\alpha)^+}$ in turn, in any order: they commute with each other. This effectively reduces us to the $\mathcal{D}_o^{\max}$ case. Now apply (1) and (2) to deduce (3) . □

COROLLARY 6.15. *Fix $\delta \in \hat{\mathcal{D}}_o^{\max}$, and suppose $\gamma \in \delta \star r_{\Omega^+}$ for some $\Omega \in \mathcal{E}_\delta$. Then Ω is uniquely determined by γ.*

Proof. By the remarks in the proof of (6.11)(3), we are reduced to the case $\delta \in \mathcal{D}_o^{\max}$. Hence we may assume that δ has the form (6.13). By the proof of (6.11)(1), we may also assume that γ has the form (6.14). (The other three possibilities, with a paired and/or b unpaired in γ, are completely analogous.) Then it is clear that the single numbers having different signs in γ and δ, along with the paired numbers in γ, uniquely determine the roots in Ω^+—they must be as shown above δ in (6.13). But Ω^+ uniquely determines Ω, by (6.10). □

Proof of Theorem E. Our strategy is to compute the radical filtration of $\pi(\delta)$ using the Kazhdan-Lusztig Conjectures (which have been proved in our context by D. Vogan [23]). However, rather than following the procedure of (2.5) (as was done in the $Sp_n\mathbb{R}$ case), we will use the "U_α-algorithm" [22, Proposition 8.13]. This permits a proof using induction on the length of δ. (Since we use the standard representations having unique irreducible quotients, whereas those in [22] have unique irreducible submodules, all the arrows must be reversed in the exact sequences.)

To set up the induction, notice that the set $\{\delta \in \mathcal{D}_o \mid n^- \in \delta\}$ together with its weak order inherited from $\mathcal{D}_o$, is naturally isomorphic (as a "labelled poset") to $\mathcal{D}_o$ for $SU(p, q-1)$: the isomorphism is given by deleting the coordinate n^-. Similar isomorphisms exist from the subsets of $\mathcal{D}_o$ involving n^+, 1^-, and 1^+ to posets for lower rank groups. For δ belonging to one of these four subsets, all the statements about $\pi(\delta)$ follow by induction on n.

The induction is started with $n = 2$, $G = SU(1,1)$. The only non-discrete series parameter in $\hat{\mathcal{D}}_o^{\max}$ is $\delta = (12)$, and $\pi(\delta)$ has radical filtration

$\overline{\pi}((12))$
$\overline{\pi}(1^+2^-) \oplus \overline{\pi}(1^-2^+)$

The corresponding collection $\mathcal{E}_\delta = \{\varnothing, \{\alpha_1\}\}$, where $\delta \star r_{\{\alpha_1\}} = \{1^+2^-, 1^-2^+\}$. The case $n = 3$ also requires a special argument. We have $G = SU(2,1)$ and the only interesting parameter is $\delta = (13)2^+$. Put $\delta' = s_2 \circ \delta = (12)3^+$. Then $\pi(\delta')$ has structure

$\overline{\pi}((12)3^+)$
$\overline{\pi}(1^+2^-3^+) \oplus \overline{\pi}(1^-2^+3^+)$

(by the $n = 2$ computation above). Applying the U_α-algorithm, $\pi(\delta)$ has structure

$\overline{\pi}((13)2^+)$
$\overline{\pi}((12)3^+) \oplus \overline{\pi}(1^+(23))$
$\overline{\pi}(1^+2^-3^+)$

.

The corresponding sets in $\mathcal{E}_\delta$ are $\varnothing, \{\alpha_2\}, \{\alpha_1\}, \text{and}\{\alpha_1, \alpha_2\}$. Thus we may assume that $n > 3$ and that both 1 and n are paired in δ.

Let us first assume that $\delta \in \mathcal{D}_o^{\max}$. Then by the remarks above, we are reduced to considering the elements of the form

$$\delta = (1\ n)\ 2^{\pm} \cdots (n-2)^{\epsilon}(n-1)^{\mu}. \tag{6.16}$$

Set $\delta' = s_{n-1} \circ \delta$; then $\delta' \prec \delta$ and

$$\delta' = (1, n-1)2^{\pm} \cdots (n-2)^{\epsilon} n^{\mu}. \tag{6.17}$$

By induction, the statements of the theorem hold for $\pi(\delta')$. Put $\alpha = \alpha_{n-1}$, $s = s_\alpha$, and notice that α is complex for δ' (and for δ). Thus the U_α-algorithm is identical to the category $\mathcal{O}$ version.

LEMMA 6.18. *Let $\delta' = (1, n-1) \cdots (n-2)^{\epsilon} n^{\mu}$. If $\lambda \in \delta' \star r_{\Lambda^+}$ for some $\Lambda \in \mathcal{E}_{\delta'}$, then either*

(a) $\lambda = \cdots (a, n-1) \cdots (n-2)^{\epsilon} n^{\mu}$, $a < n-2$,
(b) $\lambda = \cdots (n-2, n-1) n^{\mu}$, *or*
(c) $\lambda = \cdots (n-1)^{\epsilon} n^{\mu}$.

Proof. If $(e_{n-1}, \beta) = 0$ for all $\beta \in \Lambda^+$ then λ has the form (a) or (b), according to whether or not $(e_{n-2}, \beta) = 0$ for all $\beta \in \Lambda^+$. Suppose $\beta = e_i - e_{n-1} \in \Lambda^+$. If $i = n-2$ then clearly $(n-1)^{\epsilon} \in \lambda$. If $1 < i < n-2$ then the parity condition (6.8)(iv)(d) forces $i^{\epsilon} \in \delta'$, whence $(n-1)^{\epsilon} \in \lambda$. Finally if $i = 1$ then (6.9)(5) and (6.4)(ii)(a) imply that $(n-1)^{\epsilon} \in \lambda$. □

Returning to the proof of Theorem E, it will be convenient to consider two cases: $\epsilon = -\mu$ or $\epsilon = \mu$.

Case I: $\epsilon = -\mu$. Fix a composition factor $\overline{\pi}(\lambda)$ of $\pi(\delta')$, associated to $\Lambda \in \mathcal{E}_{\delta'}$ (by the induction hypothesis). Note that by (6.18), $\alpha \notin \tau(\lambda)$. We claim that every composition factor, except the top, of $\Theta_s(\overline{\pi}(\lambda))$, belongs to $\pi(\delta)$; here Θ_s is the α-wall crossing functor [22, (8.11)]. First, every composition factor $\overline{\pi}(\gamma)$ of $U_\alpha(\overline{\pi}(\lambda))$ has $\alpha \in \tau(\gamma)$, hence cannot belong to $\pi(\delta')$. Second, the socle of $\Theta_s(\pi(\lambda))$ is $\overline{\pi}(\lambda)$, which cannot belong to $\pi(\delta')$ because $\pi(\delta')$ is multiplicity free (by induction), which proves the claim. Furthermore, weight filtration considerations as in [5], and Casian's result in [3] show that if $\overline{\pi}(\lambda) \subset \text{rad}^{l}\pi(\delta')$, then $U_\alpha(\overline{\pi}(\lambda)) \subset \text{rad}^{l}\pi(\delta)$ and $\overline{\pi}(\lambda) \subset \text{rad}^{l+1}\pi(\delta)$. We are led to the following

> **Strategy:** Associate to each $\overline{\pi}(\gamma) \subset U_\alpha(\overline{\pi}(\lambda))$ a set $\Omega \in \mathcal{E}_\delta$ with $\delta \star r_{\Omega^+} = \gamma$ and $|\Omega^+| = |\Lambda^+|$, and to $\overline{\pi}(\lambda)$ a set $\Omega \in \mathcal{E}_\delta$ with $\delta \star r_{\Omega^+} = \lambda$ and $|\Omega^+| = |\Lambda^+| + 1$.

We recall that

$$U_\alpha(\overline{\pi}(\lambda)) = \overline{\pi}(s_\alpha \circ \lambda) \oplus \sum_{\gamma \in R(\lambda)} \overline{\pi}(\gamma), \tag{6.19}$$

where $R(\lambda)$ consists of all γ such that $\overline{\pi}(\gamma) \subset \text{rad}^1\pi(\lambda)$ and $\alpha \in \tau(\gamma)$. By the induction hypothesis, each such γ corresponds to a set $\Gamma \in \mathcal{E}_\lambda$ such that $\gamma \in \lambda \star r_\Gamma$ and $|\Gamma^+| = 1$. (It is important to note that γ need not be in $\mathcal{D}_o^{\max}$, but only in $\hat{\mathcal{D}}_o^{\max}$; we need the full force of the theorem.) So there are three types of composition factors $\overline{\pi}(\gamma)$ of $\Theta_s(\pi(\lambda))$ to deal with; we follow the above strategy with each type in turn.

Subcase 1: $\gamma = s_\alpha \circ \lambda$. Put $\Omega = \Lambda$. We have

$$\gamma = s_\alpha \circ \lambda = s_\alpha \circ (\delta' \star r_\Lambda) = s_\alpha \circ (s_\alpha \delta \star r_\Lambda) = \delta \star r_\Lambda = \delta \star r_\Omega$$

(as can be seen by considering the three cases in (6.18)). Obviously $|\Omega^+| = |\Lambda^+|$. We claim $\Omega \in \mathcal{E}_\delta$. For example, if $\lambda = \cdots (n-1)^\epsilon n^\mu$ and $\beta = e_i - e_{n-1} \in \Omega^+$ as in (6.18)(c), we must check that β satisfies the relevant conditions in (6.8). Note that if $i > 1$ then the parity conditions force $i^\epsilon \in \delta'$, hence $i^\epsilon \in \delta$. Since we are in the case $\epsilon = -\mu$, β is not compact for δ. (If $i = 1$ this is obvious.) In any case $\text{par}_\delta(\beta) = -\mu = \epsilon$, while $\text{par}_\delta(\zeta) = -\epsilon$ for all $\zeta \in \Omega_\beta$, so (6.8)(iv) holds. Clearly (6.8)(ii) holds with $\alpha_0 = e_1 - e_n$. The remaining conditions hold for Ω because they do for Λ.

Subcase 2: $\gamma = \lambda$. Put $\Omega = \Lambda \cup \{\alpha\}$. Then

$$\gamma = \lambda = \delta' \star r_\Lambda = (s_\alpha \circ \delta) \star r_\Lambda = (\delta \star r_\alpha) \star r_\Lambda = \delta \star r_\Omega.$$

Also $\Omega^+ = \Lambda^+ \cup \{\alpha\}$ so $|\Omega^+| = |\Lambda^+| + 1$. Again we claim $\Omega \in \mathcal{E}_\delta$. Since we already know $\Lambda \in \mathcal{E}_\delta$ (case 1), it remains only to check the conditions pertaining to α. Now the unique real roots for δ' and δ are $\alpha_0' = e_1 - e_{n-1}$ and $\alpha_0 = e_1 - e_n$, respectively. Since $\alpha_0 = \alpha_0' + \alpha_{n-1} = \alpha_0' + \alpha$, (6.8)(ii) holds for Ω. Finally, (6.8)(iv) and (v) are vacuous for $\beta = \alpha$.

Subcase 3: $\gamma \in R(\lambda)$. Suppose that $\lambda = \cdots (n-1)^\epsilon n^{-\epsilon}$ as in (6.18)(c). If $\gamma \in \lambda \star r_\Gamma$ for $\Gamma \in \mathcal{E}_\lambda$, then $\gamma = \cdots (n-1)^\epsilon n^{-\epsilon}$. In particular $\alpha \notin \tau(\gamma)$. Thus $R(\lambda) = \varnothing$.

Suppose that $\lambda = \cdots (a, n-1) \cdots (n-2)^\epsilon n^{-\epsilon}$ as in (6.18)(a). By the proof of (6.18), $(e_{n-1}, \beta) = 0$ for all $\beta \in \Lambda^+$. Now let $\gamma \in R(\lambda)$ correspond to $\Gamma \in \mathcal{E}_\lambda$.

Then $\Gamma^+ = \{\beta\}$ where $\beta = e_i - e_{n-1}$ for some $a \leq i < n-1$; otherwise $\alpha \notin \tau(\gamma)$. But then $(n-1)^\epsilon, n^{-\epsilon} \in \gamma$ and $\alpha \notin \tau(\gamma)$, a contradiction.

Suppose that $\lambda = \cdots(n-2, n-1)n^{-\epsilon}$ as in (6.18)(b). Again $(e_{n-1}, \beta) = 0$ for all $\beta \in \Lambda^+$, but $e_i - e_{n-2} \in \Lambda^+$ for some $1 \leq i < n-2$. If $\gamma \in R(\lambda)$ then $\Gamma = \{e_{n-2} - e_{n-1}\}$ and $\gamma = \cdots(n-2)^\epsilon(n-1)^{-\epsilon}n^{-\epsilon}$. (The other element γ' of $\lambda \star r_{\Gamma^+}$ has $\alpha \notin \tau(\gamma')$.) Put $\Omega = (\Lambda \setminus \{e_i - e_{n-2}\}) \cup \{e_i - e_n, e_{n-2} - e_{n-1}\}$.

CLAIM 6.20. *$\Omega \in \mathcal{E}_\delta$, $\gamma = \delta \star r_\Omega$, and $|\Omega^+| = |\Lambda^+|$.*

Proof. $\Omega^+ = (\Lambda^+ \setminus \{e_1 - e_{n-1}\}) \cup \{e_i - e_n\}$, which verifies the last statement. Evidently $(n-2)^\epsilon, (n-1)^{-\epsilon}, n^{-\epsilon}$ occur in $\delta \star r_{e_i - e_n}$ and hence also in $\delta \star r_{\Omega^+}$, while all coordinates $j < n-2$ are the same in $\delta \star r_{\Omega^+}$ as in λ and γ. Thus $\delta \star r_{\Omega^+} = \gamma$. Since $\Lambda \in \mathcal{E}_{\delta'}$ and $\mathrm{par}_{\delta'}(e_i - e_{n-2}) = -\epsilon$, all coordinates j, k between i and $n-2$ in δ' are paired in orthogonal roots $e_j - e_k$ having parity ϵ. Now $e_{n-2} - e_{n-1} \in \Omega^-$ and $\mathrm{par}_\delta(e_i - e_n) = -\epsilon$ (unless $i = 1$, in which case this latter parity is not defined). So Ω satisfies (6.8)(iv) and (v); the other conditions are easy to check. This proves the claim, and completes subcase 3. □

To complete Case I, we must show that every $\Omega \in \mathcal{E}_\delta$ has arisen precisely once in subcase 1, 2, or 3 above. Fix $\Omega \in \mathcal{E}_\delta$.

If $(e_n, \beta) = 0$ for all $\beta \in \Omega$, then $\Omega \in \mathcal{E}_{\delta'}$ and Ω occurs in case 1. If $e_{n-1} - e_n \in \Omega$ then $\Lambda = \Omega \setminus \{e_{n-1} - e_n\} \in \mathcal{E}_{\delta'}$ and Ω occurs in case 2. Finally suppose $\beta = e_i - e_n \in \Omega$ for some $i < n-1$. Then $i < n-2$ by (6.9)(4). Also $e_{n-2} - e_{n-1} \in \Omega$: if not, then $e_j - e_{n-2}, e_k - e_{n-1} \in \Omega_\beta$ for some j, k; but then $\mathrm{par}_\delta(e_j - e_{n-2}) = -\epsilon$ while $\mathrm{par}_\delta(e_k - e_{n-1}) = -\mu = \epsilon$, contradicting (6.8)(iv)(c). Set $\Lambda = (\Omega \setminus \{e_i - e_n, e_{n-2} - e_{n-1}\}) \cup \{e_i - e_{n-2}\}$. An argument similar to that given in (6.20) shows $\Lambda \in \mathcal{E}_{\delta'}$. Hence Ω occurs in case 3. It is clear from this construction that Ω arises from a unique Λ.

Case II: $\epsilon = \mu$. In this case not every composition factor $\overline{\pi}(\lambda)$ of $\pi(\delta')$ has $\alpha \notin \tau(\lambda)$. Hence if λ *does* satisfy $\alpha \notin \tau(\lambda)$, some composition factors of $U_s(\pi(\lambda))$ may belong to $\pi(\delta')$ rather than $\pi(\delta)$.

Fix λ such that $\overline{\pi}(\lambda)$ is a composition factor of $\pi(\delta')$ with $\alpha \notin \tau(\lambda)$. Then $\lambda \in \delta' \star r_{\Lambda^+}$ for some unique $\Lambda \in \mathcal{E}_{\delta'}$. Moreover since $\epsilon = \mu$, λ must be of the form $\lambda = \cdots(a, n-1)\cdots n^\mu$, $a \leq n-2$ (cf. (6.18)). We follow a similar strategy as before, except that now we must also determine whether a given composition factor of $\Theta_s(\pi(\lambda))$ belongs to $\pi(\delta')$ or $\pi(\delta)$. Let $\overline{\pi}(\gamma)$ be a composition factor of $\Theta_s(\pi(\lambda))$.

Subcase 1: $\gamma = s_\alpha \circ \lambda$. Then $\gamma = \cdots(a\ n)\cdots(n-1)^\mu$, and $\gamma \not\leq \delta'$, hence $\overline{\pi}(\gamma)$ is a composition factor of $\pi(\delta)$. It is associated to $\Omega = \Lambda$ as in Case I, subcase 1.

Subcase 2: $\gamma = \lambda$. Since $\overline{\pi}(\lambda)$ occurs with multiplicity two in $\Theta_s(\overline{\pi}(\lambda))$ and $\pi(\delta')$ has a multiplicity free composition series, the socular $\overline{\pi}(\lambda)$ belongs to $\pi(\delta)$. It is associated to $\Omega = \Lambda \cup \{\alpha\}$ as in Case I, subcase 2.

Subcase 3: $\gamma \in R(\lambda)$. Let $\gamma \in \lambda \star r_\Gamma$, $\Gamma \in \mathcal{E}_\lambda$, $|\Gamma^+| = 1$. Since $\alpha \in \tau(\gamma)$, $\Gamma^+ = \{e_i - e_{n-1}\}$ for some $a \leq i < n-1$. We consider two cases corresponding to (6.18)(b) and (a).

(i): $a = n-2$. Then $\Gamma = \{e_{n-2} - e_{n-1}\}$ and $\gamma = \cdots (n-2)^{-\epsilon}(n-1)^{\epsilon} n^{\epsilon}$. (The other element $\gamma' = \cdots (n-2)^{\epsilon}(n-1)^{-\epsilon} n^{\epsilon}$ of $\lambda \star r_{\Gamma}$ has $\alpha \notin \tau(\gamma')$.) Consider the set $\Lambda_1 = \Lambda \cup \{e_{n-2} - e_{n-1}\}$. Since $(e_{n-1}, \beta) = 0$ for all $\beta \in \Lambda$, it follows that $\Lambda_1 \in \mathcal{E}'_{\delta}$. Moreover $\delta' \star r_{\Lambda_1} = \gamma$ (not γ': apply $r_{e_{n-2}-e_{n-1}}$ first!), so that $\overline{\pi}(\gamma)$ is a composition factor of $\pi(\delta')$, in layer $|\Lambda_1^+| = |\Lambda^+| + 1$ of the radical series. This accounts for the presence of $\overline{\pi}(\gamma)$ in $\Theta_s(\overline{\pi}(\lambda))$.

(ii): $a < n-2$. Here $\lambda = \cdots (a, n-1) \cdots (n-2)^{\epsilon} n^{\epsilon}$; recall that $\Gamma^+ = \{e_i - e_{n-1}\}$. If $i > a$ then the parity condition forces $i^{\epsilon} \in \lambda$, hence also $i^{\epsilon} \in \delta'$. Then $\Lambda_1 = \Lambda \cup \Gamma \in \mathcal{E}_{\delta'}$ and $\gamma = \delta' \star r_{\Lambda_1}$. Thus $\overline{\pi}(\gamma)$ belongs to $\pi(\delta')$ again. Suppose $i = a$. Then $\gamma = \cdots a^{-\epsilon} \cdots (n-2)^{\epsilon}(n-1)^{\epsilon} n^{\epsilon}$, and the construction in the previous paragraph will show that $\overline{\pi}(\gamma)$ belongs to $\pi(\delta')$, *provided* that $a^{\epsilon} \in \delta'$ or $a = 1$. (In the last case $\lambda = \delta'$.) Assume that $a^{-\epsilon} \in \delta'$; i.e. $\delta' = (1, n-1) \cdots a^{-\epsilon} \cdots (n-2)^{\epsilon} n^{\epsilon}$. Then $e_a - e_{n-1} \notin \Lambda_1$ for any $\Lambda_1 \in \mathcal{E}_{\delta'}$, so $\overline{\pi}(\gamma)$ must be a composition factor of $\pi(\delta)$. Now Λ^+ contains an element $e_i - e_a$ having parity ϵ for δ'. Put $\Omega = (\Lambda \setminus \{e_i - e_a\}) \cup \Gamma \cup \{e_i - e_n\}$. Notice that $\Omega^+ = (\Lambda^+ \setminus \{e_i - e_a\}) \cup \{e_i - e_n\}$, so $|\Omega^+| = |\Lambda^+|$. If we set $\beta = e_i - e_n$, then

$$\Omega_{\beta} = \Lambda_{e_i - e_a} \cup \Gamma_{e_a - e_{n-1}} \cup \{e_a - e_{n-1}\}, \tag{6.21}$$

a disjoint union of mutually orthogonal roots having parity $-\epsilon$ for δ and "involving" every e_j, $i < j < n$. Also if $i > 1$, $\mathrm{par}_{\delta}(\beta) = \epsilon$. It follows that $\Omega \in \mathcal{E}_{\delta}$. Since

$$\begin{aligned} \delta \star r_{\beta} &= \begin{cases} 1^{-\epsilon} \cdots a^{-\epsilon} \cdots (n-2)^{\epsilon}(n-1)^{\epsilon} n^{\epsilon}, & i = 1 \\ (1\ i) \cdots a^{-\epsilon} \cdots (n-2)^{\epsilon}(n-1)^{\epsilon} n^{\epsilon}, & i > 1 \end{cases} \\ &= (\delta' \star r_{e_i - e_a}) \star r_{e_a - e_{n-1}}, \end{aligned}$$

it follows that $\delta \star r_{\Omega} = \gamma$. This completes (ii) and hence subcase 3.

To complete Case II, we must show exhaustion of $\mathcal{E}_{\delta}$. To this end, fix $\Omega \in \mathcal{E}_{\delta}$. If $(e_n, \beta) = (e_{n-1}, \beta) = 0$ for all $\beta \in \Omega$, then $\Omega \in \mathcal{E}_{\delta'}$, and $\lambda = \delta' \star r_{\Omega}$ has $\alpha \notin \tau(\lambda)$; thus Ω occurs in subcase 1. If $e_{n-1} - e_n \in \Omega$, then $\Lambda = \Omega - \{e_{n-1} - e_n\} \in \mathcal{E}_{\delta'}$ and Ω occurs in subcase 2. Finally, observe that $e_j - e_{n-1} \notin \Omega^+$ for any $j < n-1$, since $\mathrm{sgn}_{\delta}(n-1) = \mathrm{sgn}_{\delta}(n-2)$. So the only remaining possibility is $\beta = e_i - e_n \in \Omega^+$ for some $i < n-2$, and $e_a - e_{n-1} \in \Omega_{\beta} \subset \Omega$ for some $a > i$. By (6.9)(3), Ω_{β} splits up into a disjoint union of the form (6.21), and therefore Ω occurs in subcase 3. It is again clear from the construction that Ω arises from a unique Λ.

This completes the proof of statements (1) and (3) of Theorem E for the parameters $\delta \in \mathcal{D}_o^{\max}$.

We now treat the general case $\delta \in \hat{\mathcal{D}}_o^{\max}$. According to (6.2), $\delta = \cdots (a_1 b_1) \cdots \cdots (a_r b_r) \cdots$, with $a_1 < b_1 < \cdots < a_r < b_r$. By the induction hypothesis, we may assume that $b_r = n$. The argument given for the $\mathcal{D}_o^{\max}$ case can then be repeated, the only change being to replace the coordinate 1 with a_r; keep in mind the decomposition of sets $\Omega \in \mathcal{E}_{\delta}$ as in the proof of (6.11)(3). We leave the details to the reader.

The multiplicity one statement, (2), is Corollary 6.15.

Finally, to prove (4), recall that a composition factor $\overline{\pi}(\gamma) \subset \mathrm{rad}^i \pi(\delta)$ is called *standard* if $i = \ell^I(\delta) - \ell^I(\gamma)$. Let $\gamma \in \delta \star r_{\Omega^+}$ with $\Omega \in \mathcal{E}_\delta$ and $|\Omega^+| = i$. Observe that, for $\beta \in \Omega^+$, $\ell^I(\delta) - \ell^I(\delta \star r_\beta) \geq 1$, with equality if and only if β is simple. By the Heriditarity Lemma (6.12) and induction on i, $\overline{\pi}(\gamma)$ is standard if and only if every root in Ω^+ is simple. By (6.8)(iii) and (6.9)(1), this is equivalent to the condition $\Omega^- = \varnothing$, i.e. $\Omega = \Omega^+$. □

COROLLARY 6.22. *Let $G = SO^*(2n)$. Then for each pyramid $\mathcal{P}_{\sigma,M} \subset \mathcal{D}_o$ there exists a subset $\mathcal{T}_\sigma \subset \mathcal{P}_{\sigma,M}$ with the following properties:*

(i) $|\mathcal{T}_\sigma| = \frac{1}{2}|\mathcal{P}_{\sigma,M}|$;
(ii) *there exists $\delta_\sigma \in \mathcal{T}_\sigma$ such that $\ell^I(\delta_\sigma) > \frac{1}{2}\sup\{\ell^I(\delta) \mid \delta \in \mathcal{P}_{\sigma,M}\}$;*
(iii) *$\pi(\delta)$ is multiplicity free for all $\delta \in \mathcal{T}_\sigma$.*

Proof. Recall from (2.11) the parametrization of an element $\delta \in \mathcal{D}_o^{\max}$ for $SO^*(2n)$, as an arrangement of the numbers $1, \ldots, n$ into singles and pairs; $\mathcal{D}_o^{\max}$ consists of those parameters with exactly one pair. Each pyramid has the form $\mathcal{P}_{\sigma,M} = \{\delta \in \mathcal{D}_o^{\max} \mid \delta \preceq \delta_{\sigma,M}\}$, where $\delta_{\sigma,M} = 1^{\epsilon_1} \cdots (n-2)^{\epsilon_{n-2}}(n-1,n)^b$ for some choice (depending on σ) of $\epsilon_1, \ldots, \epsilon_{n-2} \in \{+,-\}$.

Set $\delta_\sigma = (1,n)^\sharp 2^{-\epsilon_1} 3^{\epsilon_2} \cdots (n-1)^{\epsilon_{n-2}}$, $\mathcal{T}_\sigma = \{\delta \in \mathcal{D}_o^{\max} \mid \delta \preceq \delta_\sigma\}$. Then $\mathcal{T}_\sigma \subset \mathcal{P}_{\sigma,M}$, and it is an easy combinatorial exercise, using (2.6), to check (i) and (ii). Moreover, $\hat{\mathcal{T}}_\sigma = \{\delta \in \mathcal{D}_o \mid \delta \preceq \delta_\sigma\}$ is isomorphic, as a labelled poset, to some G-pyramid for some $SU(p,q)$ (depending on σ) with $p+q = n$. Hence (iii) follows from Theorem E. □

EXAMPLE. We now illustrate the ideas in this chapter via the case $G = SU(3,2)$. Consider the parameter $\delta = (15)23^-4$ (omitting superscript $+$ signs for brevity), which is the top of a non-holomorphic pyramid $\mathcal{P}_{\sigma,M}$. This is the most complicated pyramid, owing to the fact that the corresponding generalized principal series representations are induced from large discrete series representations of $SU(2,1)$. In Table 6.1, we tabulate the data necessary to apply Theorem E. We also give the radical filtration of $\pi((15)23^-4)$. Let's isolate three specific calculations.

The first two examples illustrate the delicate nature of (6.6). Begin with the case $\Omega = \{e_1 - e_2, e_2 - e_3\} = \Omega^+$. Then

$$\begin{aligned} \delta \star r_{\Omega^+} &= (\delta \star r_{\alpha_1}) \star r_{\alpha_2} \cap (\delta \star r_{\alpha_2}) \star r_{\alpha_1} \\ &= \{12^-(35)4\} \cap \{12^-(35)4,\ 1^-2(35)4\} \\ &= \{12^-(35)4\}. \end{aligned}$$

Since $|\Omega^+| = 2$, $\overline{\pi}(12^-(35)4)$ will lie in the 2-layer of the radical filtration of $\pi((15)23^-4)$; keep in mind that the top layer is the 0-layer. Similarly, consider the case when $\Omega = \{ e_1 - e_2, e_2 - e_3, e_3 - e_4 \} = \Omega^+$. Then

$$\begin{aligned}\delta \star r_{\Omega^+} &= \cap_{\sigma \in S_3} \big((\delta \star r_{\alpha_{\sigma(1)}}) \star r_{\alpha_{\sigma(2)}} \big) \star r_{\alpha_{\sigma(3)}} \\ &= \{ 12^-3(45) \} \cap \{ 12^-3(45), 123^-(45) \} \cap \{ 12^-3(45), 1^-23(45) \} \cap \\ &\quad \cap \{ 1^-23(45), 12^-3(45) \} \cap \{ 12^-3(45), 123^-(45) \} \cap \\ &\quad \cap \{ 12^-3(45), 1^-23(45) \} \\ &= \{ 12^-3(45) \}.\end{aligned}$$

Since $|\Omega^+| = 3$, $\overline{\pi}(12^-3(45))$ will lie in the 3-layer of the radical filtration of $\pi((15)23^-4)$. The last calculation illustrates the subtle nature of (6.4). Consider $\Omega = \{ e_1 - e_2, e_2 - e_5, e_3 - e_4 \}$, $\Omega^+ = \{ e_1 - e_2, e_2 - e_5 \}$. Then

$$\begin{aligned}\delta \star r_{\Omega^+} &= (\delta \star r_{e_1 - e_2}) \star r_{e_2 - e_5} \cap (\delta \star r_{e_2 - e_5}) \star r_{e_1 - e_2} \\ &= \{ 12^-3^-45 \} \cap \{ 1^-23^-45, 12^-3^-45 \} \\ &= \{ 12^-3^-45 \}.\end{aligned}$$

Since $|\Omega^+| = 2$, $\overline{\pi}(12^-3^-45)$ will lie in the 2-layer of the radical filtration of $\pi((15)23^-4)$.

Of course, it must be emphasized that in practice, the computation of $\delta \star r_{\Omega^+}$ is greatly simplified by the proof of (6.11), particularly (6.13) and (6.14), which completely eliminates the need to use permuations.

The structure of the radical filtration of $\pi((15)23^-4)$ is given as follows:

$(15)23^-4$
$(14)23^-5 \oplus 1(25)3^-4 \oplus (12)(35)4 \oplus (13)2(45) \oplus (12)3^-45 \oplus 123^-(45)$
$(13)24^-5 \oplus 1(24)3^-5 \oplus (12)(34)5 \oplus 12^-(35)4 \oplus 1(23)(45)$ $\oplus (12)3(45) \oplus 123^-4^-5 \oplus 12^-3^-45$
$1(23)4^-5 \oplus (12)34^-5 \oplus 12^-(34)5 \oplus 12^-3(45)$
12^-34^-5

TABLE 6.1. Theorem E data for $\delta_{(15)23^-4} \in \mathcal{D}_{SU(3,2)}$

Ω	Ω^+	$\lvert\Omega^+\rvert$	$\delta \star r_{\Omega^+}$
$\varnothing$	$\varnothing$	0	$(15)23^-4$
$e_1 - e_2$	$e_1 - e_2$	1	$1(25)3^-4$
$e_2 - e_3$	$e_2 - e_3$	1	$(12)(35)4$
$e_3 - e_4$	$e_3 - e_4$	1	$(13)2(45)$
$e_4 - e_5$	$e_4 - e_5$	1	$(14)23^-5$
$e_1 - e_2,\ e_2 - e_3$	$e_1 - e_2,\ e_2 - e_3$	2	$12^-(35)4$
$e_1 - e_2,\ e_3 - e_4$	$e_1 - e_2,\ e_3 - e_4$	2	$1(23)(45)$
$e_1 - e_2,\ e_4 - e_5$	$e_1 - e_2,\ e_4 - e_5$	2	$1(24)3^-5$
$e_2 - e_3,\ e_3 - e_4$	$e_2 - e_3,\ e_3 - e_4$	2	$(12)3(45)$
$e_2 - e_3,\ e_4 - e_5$	$e_2 - e_3,\ e_4 - e_5$	2	$(12)(34)5$
$e_3 - e_4,\ e_4 - e_5$	$e_3 - e_4,\ e_4 - e_5$	2	$(13)24^-5$
$e_1 - e_2,\ e_2 - e_3,\ e_3 - e_4$	$e_1 - e_2,\ e_2 - e_3,\ e_3 - e_4$	3	$12^-3(45)$
$e_1 - e_2,\ e_2 - e_3,\ e_4 - e_5$	$e_1 - e_2,\ e_2 - e_3,\ e_4 - e_5$	3	$12^-(34)5$
$e_1 - e_2,\ e_3 - e_4,\ e_4 - e_5$	$e_1 - e_2,\ e_3 - e_4,\ e_4 - e_5$	3	$1(23)4^-5$
$e_2 - e_3,\ e_3 - e_4,\ e_4 - e_5$	$e_2 - e_3,\ e_3 - e_4,\ e_4 - e_5$	3	$(12)34^-5$
$e_1 - e_2,\ e_2 - e_3,$ $e_3 - e_4,\ e_4 - e_5$	$e_1 - e_2,\ e_2 - e_3,$ $e_3 - e_4,\ e_4 - e_5$	4	12^-34^-5
$e_1 - e_4,\ e_2 - e_3$	$e_1 - e_4$	1	$123^-(45)$
$e_2 - e_5,\ e_3 - e_4$	$e_2 - e_5$	1	$(12)3^-45$
$e_1 - e_2,\ e_2 - e_5,\ e_3 - e_4$	$e_1 - e_2,\ e_2 - e_5$	2	12^-3^-45
$e_1 - e_4,\ e_2 - e_3,\ e_4 - e_5$	$e_1 - e_4,\ e_4 - e_5$	2	123^-4^-5

7. THE EXCEPTIONAL CASES

In this chapter, we describe how one obtains explicit formulas for the radical filtration of $\pi(\delta)$, $\delta \in \mathcal{D}^{\max}$, whenever G is of exceptional type. An obvious approach would be to simply list the data for this case. However, we offer an alternate tack, which has the added advantage of suggesting a general conjecture for any G of Hermitian symmetric type.

In the setting of Theorem 4.8, which assumes holomorphic type inducing data, we showed that the composition factors of $\pi(\delta)$ can be predicted using the Enright-Shelton theory. One aspect of Theorem 4.8 may be interpreted as saying:

(7.1) For $\delta \in \mathcal{P}_{\text{holo},M}$, the Enright-Shelton theory predicts the composition factors of $\pi(\delta)$, "up to $\mathbf{Q}^\natural$-orbit".

Our next result asserts that the qualitative flavor of (7.1) holds in general, in the exceptional cases; in turn, this suggests a conjecture for any G of Hermitian symmetric type.

THEOREM F. *Let G be of type HS.6 or HS.7 and $\delta \in \mathcal{D}^{\max}$. Then $\overline{\pi}(\gamma)$ is a composition factor of $\pi(\delta)$ only if $\mathbb{A}(\gamma_o) = \overline{\mathbb{A}(\delta_o) r_\Omega}$, for some $\Omega \in \mathcal{E}_{\mathbb{A}(\delta_o)}$.*

CONJECTURE 7.2. *Let G be of Hermitian symmetric type and $\delta \in \mathcal{D}_P^{\max}$, where P satisfies the conditions in Theorem D. Then $\overline{\pi}(\gamma)$ is a composition factor of $\pi(\delta)$ only if $\mathbb{A}(\gamma_o) = \overline{\mathbb{A}(\delta_o) r_\Omega}$, for some $\Omega \in \mathcal{E}_{\mathbb{A}(\delta_o)}$.*

Beyond the cases covered by Theorems 4.8 and F, the conjecture has been verified for $SU(p,q), 1 \leq q \leq p \leq 3$.

For the remainder of this chapter, we assume $G = E_{6,-14}$; the case of $E_{7,-25}$ can be handled in a similar way, using the data in Table 9.2. Note that $\mathcal{D} = V$, so there is no confusion in writing δ in place of δ_o. Up to conjugacy, there is a unique maximal cuspidal parabolic subgroup $P = MAN$, which has $M_{ss} = SU(5,1)$. As discussed in (4.23),

(7.3)
$$\mathcal{D}^{\max} = \bigcup_{1 \leq i \leq 6} \mathcal{P}_{\sigma_i, M},$$

where $\mathcal{P}_{\sigma_6,M}$ denotes the holomorphic pyramid. Using Table 9.1, one can verify poset isomorphisms:

(7.4)
$$\hat{\mathcal{P}}_{\sigma_6,M} = \hat{\mathcal{P}}_{\sigma_1,M}\,;\ \hat{\mathcal{P}}_{\sigma_5,M} = \hat{\mathcal{P}}_{\sigma_2,M}\,;\ \hat{\mathcal{P}}_{\sigma_4,M} = \hat{\mathcal{P}}_{\sigma_3,M}.$$

For this reason, we need only study three of these pyramids, one of which is holomorphic.

In (4.23b), we parametrized $\mathcal{P}_{\mathrm{holo},M}$, via a set $\mathcal{X}_6$. Similarly, using the notation of Table 9.1, we have $\mathcal{P}_{\sigma_5,M} = \{\delta_i \mid i \in \mathcal{X}_5\}$ and $\mathcal{P}_{\sigma_4,M} = \{\delta_i \mid i \in \mathcal{X}_4\}$, where

$$\begin{aligned}(7.5)\qquad \mathcal{X}_5 = \{&46,50,53,54,56,57,86,91,94,95,97,130,136,137,\\&140,141,180,181,188,189,191,234,235,241,242,\\&286,287,292,335,336,341,381,382,420,450,475\}\end{aligned}$$

$$\begin{aligned}(7.6)\qquad \mathcal{X}_4 = \{&43,48,49,51,52,55,83,88,89,90,93,127,132,133,\\&134,135,176,177,184,185,186,230,231,238,239,\\&283,284,290,332,333,339,378,379,418,449,474\}\end{aligned}$$

Recall from (4.23) the set $\mathcal{S}$ which indexes a portion of the diagram $W^{\mathbf{Q}^\sharp}$ in [10, Fig. 7.1]; again, we remind the reader that one must convert the Dynkin diagram labeling of [10] to our conventions in (2.17). Apply the theory of [9] and adopt the conventions in (4.17) to arrive at the following $\mathbf{K}$-orbit decompositions of the $\mathbf{Q}^\sharp$-orbits indexed by $\mathcal{S}$.

LEMMA 7.7. *We have the following equalities in $\mathcal{M}_{\mathrm{orb}}$ (recall (4.17)):*

$$\begin{aligned}
w_{0^*} &= \delta_{27};\\
w_{1^*} &= \delta_{63} + t\delta_{26};\\
w_{2^*} &= \delta_{103} + t\delta_{62} + t^2\delta_{25};\\
w_{3^*} &= \delta_{148} + t\delta_{102} + t^2\delta_{61} + t^3\delta_{24};\\
w_{4^*} &= \delta_{199} + t\delta_{147} + t^2\delta_{101} + t^3\delta_{60} + t^4\delta_{23};\\
w_{5^*} &= \delta_{198} + t\delta_{146} + t^2\delta_{100} + t^3\delta_{59} + t^4\delta_{22};\\
w_{6^*} &= \delta_{251} + t\delta_{197} + t^2\delta_{145} + t^3\delta_{98} + t^4(\delta_{56} + \delta_{57}) + t^5\delta_{20};\\
w_{7^*} &= \delta_{249} + t\delta_{196} + t^2\delta_{144} + t^3\delta_{99} + t^4\delta_{58} + t^5\delta_{21};\\
w_{8^*} &= \delta_{300} + t\delta_{250} + t^2\delta_{192} + t^3(\delta_{140} + \delta_{143}) + t^4(\delta_{94} + \delta_{95}) + t^5\delta_{53} + t^6\delta_{18};\\
w_{9^*} &= \delta_{299} + t\delta_{248} + t^2\delta_{195} + t^3\delta_{142} + t^4(\delta_{96} + \delta_{97}) + t^5(\delta_{54} + \delta_{55}) + t^6\delta_{19};\\
w_{10^*} &= \delta_{348} + t\delta_{293} + t^2(\delta_{242} + \delta_{245}) + t^3(\delta_{188} + \delta_{194}) + t^4(\delta_{136} + \delta_{137})\\
&\quad + t^5\delta_{91} + t^6\delta_{50} + t^7\delta_{16};\\
w_{11^*} &= \delta_{347} + t\delta_{298} + t^2\delta_{244} + t^3(\delta_{191} + \delta_{193}) + t^4(\delta_{138} + \delta_{141}) + t^5(\delta_{92} + \delta_{93})\\
&\quad + t^6(\delta_{51} + \delta_{52}) + t^7\delta_{17};\\
w_{12^*} &= \delta_{383} + t(\delta_{336} + \delta_{343}) + t^2(\delta_{287} + \delta_{296}) + t^3(\delta_{234} + \delta_{247})\\
&\quad + t^4(\delta_{180} + \delta_{181}) + t^5\delta_{130} + t^6\delta_{86} + t^7\delta_{46} + t^8\delta_{13};\\
w_{13^*} &= \delta_{391} + t\delta_{342} + t^2(\delta_{292} + \delta_{295}) + t^3(\delta_{241} + \delta_{246}) + t^4(\delta_{182} + \delta_{189})\\
&\quad + t^5(\delta_{131} + \delta_{132}) + t^6(\delta_{87} + \delta_{88}) + t^7(\delta_{47} + \delta_{48}) + t^8\delta_{14};
\end{aligned}$$

$$\begin{aligned}
w_{14^*} =& \delta_{390} + t\delta_{346} + t^2\delta_{291} + t^3(\delta_{240} + \delta_{243}) + t^4(\delta_{186} + \delta_{187} + \delta_{190}) \\
&+ t^5(\delta_{134} + \delta_{135} + \delta_{139}) + t^6(\delta_{89} + \delta_{90}) + t^7\delta_{49} + t^8\delta_{15}; \\
w_{15^*} =& \delta_{421} + t(\delta_{382} + \delta_{389}) + t^2(\delta_{335} + \delta_{345}) + t^3(\delta_{286} + \delta_{297}) + t^4(\delta_{228} + \delta_{235}) \\
&+ t^5(\delta_{175} + \delta_{176}) + t^6(\delta_{126} + \delta_{127}) + t^7(\delta_{82} + \delta_{83}) + t^8(\delta_{42} + \delta_{43}) + t^9\delta_{11}; \\
w_{16^*} =& \delta_{425} + t\delta_{388} + t^2(\delta_{340} + \delta_{341}) + t^3(\delta_{288} + \delta_{294}) + t^4(\delta_{236} + \delta_{237} + \delta_{238}) \\
&+ t^5(\delta_{183} + \delta_{184} + \delta_{185}) + t^6(\delta_{128} + \delta_{133}) + t^7(\delta_{84} + \delta_{85}) \\
&+ t^8(\delta_{44} + \delta_{45}) + t^9\delta_{12}; \\
w_{17^*} =& \delta_{451} + t(\delta_{420} + \delta_{424}) + t^2(\delta_{381} + \delta_{387}) + t^3(\delta_{330} + \delta_{344}) \\
&+ t^4(\delta_{281} + \delta_{282} + \delta_{283}) + t^5(\delta_{229} + \delta_{230} + \delta_{231}) + t^6(\delta_{170} + \delta_{177}) \\
&+ t^7(\delta_{121} + \delta_{122}) + t^8(\delta_{77} + \delta_{78} + \delta_{79}) + +t^9(\delta_{39} + \delta_{40}) + t^{10}\delta_9; \\
w_{18^*} =& \delta_{453} + t\delta_{422} + t^2(\delta_{384} + \delta_{385}) + t^3(\delta_{337} + \delta_{338} + \delta_{339}) \\
&+ t^4(\delta_{280} + \delta_{289} + \delta_{290}) + t^5(\delta_{226} + \delta_{233} + \delta_{239}) + t^6(\delta_{173} + \delta_{174} + \delta_{178}) \\
&+ t^7(\delta_{124} + \delta_{125} + \delta_{129}) + t^8(\delta_{80} + \delta_{81}) + t^9\delta_{41} + t^{10}\delta_{10}; \\
w_{19^*} =& \delta_{476} + t(\delta_{450} + \delta_{452}) + t^2(\delta_{415} + \delta_{423}) + t^3(\delta_{376} + \delta_{378} + \delta_{386}) \\
&+ t^4(\delta_{327} + \delta_{331} + \delta_{332}) + t^5(\delta_{276} + \delta_{278} + \delta_{284}) + t^6(\delta_{222} + \delta_{224} + \delta_{225}) \\
&+ t^7(\delta_{165} + \delta_{171} + \delta_{172}) + t^8(\delta_{116} + \delta_{117} + \delta_{123}) + t^9(\delta_{74} + \delta_{75}) \\
&+ t^{10}(\delta_{36} + \delta_{37}) + t^{11}\delta_7; \\
w_{20^*} =& \delta_{477} + t\delta_{440} + t^2(\delta_{408} + \delta_{414}) + t^3(\delta_{368} + \delta_{375} + \delta_{380}) \\
&+ t^4(\delta_{321} + \delta_{322} + \delta_{329} + \delta_{334}) + t^5(\delta_{271} + \delta_{273} + \delta_{279} + \delta_{285}) \\
&+ t^6(\delta_{219} + \delta_{220} + \delta_{227} + \delta_{232}) + t^7(\delta_{167} + \delta_{168} + \delta_{179}) \\
&+ t^8(\delta_{119} + \delta_{120}) + t^9\delta_{76} + t^{10}\delta_{38} + t^{11}\delta_8; \\
w_{21^*} =& \delta_{492} + t(\delta_{472} + \delta_{473}) + t^2(\delta_{447} + \delta_{448} + \delta_{449}) + t^3(\delta_{416} + \delta_{417} + \delta_{418}) \\
&+ t^4(\delta_{364} + \delta_{377} + \delta_{379}) + t^5(\delta_{315} + \delta_{320} + \delta_{333}) + t^6(\delta_{265} + \delta_{266} + \delta_{269}) \\
&+ t^7(\delta_{213} + \delta_{214} + \delta_{217} + \delta_{218}) + t^8(\delta_{159} + \delta_{160} + \delta_{161} + \delta_{166}) \\
&+ t^9(\delta_{112} + \delta_{113} + \delta_{118}) + t^{10}(\delta_{70} + \delta_{71}) + t^{11}\delta_{33} + t^{12}\delta_5; \\
w_{22^*} =& \delta_{493} + t(\delta_{467} + \delta_{475}) + t^2(\delta_{441} + \delta_{446}) + t^3(\delta_{409} + \delta_{412} + \delta_{419}) \\
&+ t^4(\delta_{369} + \delta_{370} + \delta_{373} + \delta_{374}) + t^5(\delta_{323} + \delta_{324} + \delta_{326} + \delta_{328}) \\
&+ t^6(\delta_{272} + \delta_{274} + \delta_{275} + \delta_{277}) + t^7(\delta_{215} + \delta_{221} + \delta_{223}) \\
&+ t^8(\delta_{162} + \delta_{163} + \delta_{169}) + t^9(\delta_{114} + \delta_{115}) + t^{10}(\delta_{72} + \delta_{73}) \\
&+ t^{11}(\delta_{34} + \delta_{35}) + t^{12}\delta_6.
\end{aligned}$$

It remains to exhibit an algorithm, in the spirit of (1.3)–(1.5), which describes the radical filtration of $\pi(\delta)$, $\delta \in \mathcal{P}_{\sigma_6,M} \cup \mathcal{P}_{\sigma_5,M} \cup \mathcal{P}_{\sigma_4,M}$. To motivate the type of result we want, we urge the reader to review Theorem 4.8, which handles the holomorphic pyramid case $\mathcal{P}_{\sigma_6,M}$. The setting of Theorem 4.8 is most manageable, due to the fact we have associated *pairs* to each Enright-Shelton set

$\Omega \in \mathcal{E}_{\mathbb{A}(\delta)}$. Geometrically, this corresponds to the properties in Lemma 4.12. To handle the other two pyramids we will need a substitute for this lemma. Though we will not need to speak in these terms, this will ultimately require a closer analysis of the difference between the intersection cohomology of the closure of $\mathcal{V}_\gamma$ in $\mathcal{O}_{\mathbb{A}(\gamma)}$ and in $\mathcal{B}$.

Let $\delta \in \mathcal{P}_{\sigma_4,M} \cup \mathcal{P}_{\sigma_5,M} \cup \mathcal{P}_{\sigma_6,M}$. Recall the discussion in (1.3)–(1.5). We define $\mathcal{E}_\delta = \mathcal{E}_{\mathbb{A}(\delta)}$ and now describe a map $\mathbb{E}_\delta : \mathcal{E}_\delta \to \mathcal{M}_{\mathbf{K}}$. Our description of the map will produce an algorithm which yields closed formulas for the Laurent polynomials $E_{\gamma,\delta,\Omega}$.

The first step in describing $\mathbb{E}_\delta$ is to associate a poset $\mathcal{T}_w$ to each $\mathbf{Q}^\sharp$-orbit $\mathcal{O}_w$. The elements of $\mathcal{T}_w$ are the parameters in V arising in the $\mathcal{M}_{\mathrm{orb}}$ expressions of Lemma 7.7. Furthermore, the codimension function provides a notion of length for each element of $\mathcal{T}_w$, partitioning the set into "levels". Introduce the order relations coming from $\preceq$ on V. This means we can attach labels in the set $\{1,2,3,4,5,6\}$ (corresponding to the indexing in the Satake diagram (2.12a)) to the closure weak order relations. Finally, an order relation in the strong order but *not* in the weak order can acquire a numerical label if it arises via a Type I Cayley transform. More precisely, suppose

(7.8)
$$\begin{aligned}
&\delta, \gamma \in \mathcal{T}_w;\\
&s \text{ is of noncompact Type I for } \delta;\\
&\gamma \preceq \delta, \gamma \preceq s \times \delta;\\
&\gamma \not\preceq_w \delta\\
&\gamma \not\preceq_w s \times \delta\\
&\ell^I(\gamma) = \ell^I(\delta) - 1 = \ell^I(s \times \delta) - 1.
\end{aligned}$$

In the scenario (7.8), we introduce a "fake s" labeling of the closure relation relating $\gamma \preceq \delta$ and $\gamma \preceq s \times \delta$ as in the figure below:

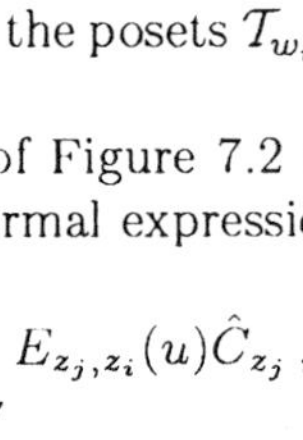

As examples, Figure 7.1 indicates the posets $\mathcal{T}_{w_{i^*}}$ attached to the orbits $\mathcal{O}_{w_{i^*}}$, $i =$ 16, 14, 13, 11, 9.

Refer to the labeled poset $\mathcal{C}$ of Figure 7.2 and attach to each node "i" the parameter "z_i". Then we have formal expressions

(7.9)
$$z_i = \sum_{1 \le j \le 17} E_{z_j,z_i}(u)\hat{C}_{z_j}, \; 1 \le i \le 17$$

in the module $\mathcal{M}_{\mathrm{formal}} = \mathbb{Z}[u^{1/2}, u^{-1/2}] \otimes_{\mathbb{Z}} \mathbb{Z}[\{\hat{C}_{z_j} \mid 1 \le j \le 17\}]$ by identifying this figure with an appropriately relabeled truncated poset of [10, Fig. 7.1]. In so doing, we find the following expressions in $\mathcal{M}_{\text{formal}}$:

$$
\begin{aligned}
(7.10) \qquad z_1 &= \hat{C}_{z_1} - u^{-1/2}\hat{C}_{z_2} + u^{-1}\hat{C}_{z_{13}};\\
z_2 &= \hat{C}_{z_2} - u^{-1/2}[\hat{C}_{z_3} + \hat{C}_{z_{13}}] + u^{-1}\hat{C}_{z_{11}};\\
z_3 &= \hat{C}_{z_3} - u^{-1/2}[\hat{C}_{z_4} + \hat{C}_{z_{11}}] + u^{-1}\hat{C}_{z_9};\\
z_4 &= \hat{C}_{z_4} - u^{-1/2}[\hat{C}_{z_5} + \hat{C}_{z_6} + \hat{C}_{z_9}] + u^{-1}\hat{C}_{z_7};\\
z_5 &= \hat{C}_{z_5} - u^{-1/2}\hat{C}_{z_7} + u^{-1}\hat{C}_{z_{15}};\\
z_6 &= \hat{C}_{z_6} - u^{-1/2}[\hat{C}_{z_7} + \hat{C}_{z_8}] + u^{-1}\hat{C}_{z_{10}};\\
z_7 &= \hat{C}_{z_7} - u^{-1/2}[\hat{C}_{z_9} + \hat{C}_{z_{10}} + \hat{C}_{z_{15}}] + u^{-1}\hat{C}_{z_{12}};\\
z_8 &= \hat{C}_{z_8} - u^{-1/2}\hat{C}_{z_{10}};\\
z_9 &= \hat{C}_{z_9} - u^{-1/2}[\hat{C}_{z_{11}} + \hat{C}_{z_{12}}] + u^{-1}\hat{C}_{z_{14}};\\
z_{10} &= \hat{C}_{z_{10}} - u^{-1/2}\hat{C}_{z_{12}};\\
z_{11} &= \hat{C}_{z_{11}} - u^{-1/2}[\hat{C}_{z_{13}} + \hat{C}_{z_{14}}] + u^{-1}\hat{C}_{z_{16}};\\
z_{12} &= \hat{C}_{z_{12}} - u^{-1/2}[\hat{C}_{z_{14}} + \hat{C}_{z_{15}}] + u^{-1}\hat{C}_{z_{17}};\\
z_{13} &= \hat{C}_{z_{13}} - u^{-1/2}\hat{C}_{z_{16}};\\
z_{14} &= \hat{C}_{z_{14}} - u^{-1/2}[\hat{C}_{z_{16}} + \hat{C}_{z_{17}}];\\
z_{15} &= \hat{C}_{z_{15}} - u^{-1/2}\hat{C}_{z_{17}};\\
z_{16} &= \hat{C}_{z_{16}};\\
z_{17} &= \hat{C}_{z_{17}}.
\end{aligned}
$$

Philosophically, one might view these as intersection cohomology decompositions in some kind of "truncated category."

We are now ready to describe an algorithm to obtain the expressions in (1.5). We fix $\delta \in \mathcal{P}_{\sigma_i, M}$, $i = 4, 5, 6$, and put $\mathcal{E}_\delta = \mathcal{E}_{\mathbb{A}(\delta)}$.

Step 1: For each $\Omega \in \mathcal{E}_\delta$, determine the labeled poset $\mathcal{T}_{\overline{\mathbb{A}(\delta)r_\Omega}}$.

Step 2: For each $\Omega \in \mathcal{E}_\delta$, denote by $\mathcal{C}(\Omega)$ the largest subset of $\mathcal{C}$ containing z_1 and having the property of being labeled poset isomorphic to a subset (denoted $\mathcal{C}_\Omega$) of $\mathcal{T}_{\overline{\mathbb{A}(\delta)r_\Omega}}$ which contains $\mathbb{B}(\overline{\mathbb{A}(\delta)r_\Omega})$. In other words, we superimpose the largest possible portion of $\mathcal{C}$ onto the top of $\mathcal{T}_{\overline{\mathbb{A}(\delta)r_\Omega}}$. Let $\psi_\Omega : \mathcal{C}_\Omega \to \mathcal{C}(\Omega)$ be the bijection underlying this superimposition process.

Step 3: For each $\Omega \in \mathcal{E}_\delta$, we define Laurent polynomials in $\mathbb{Z}[u^{1/2}, u^{-1/2}]$ as follows:

$$E_{\gamma,\delta,\Omega}(u) = \begin{cases} E_{\psi_\Omega(\gamma),\psi_{\varnothing}(\delta)}(u), & \text{if } \gamma \in \mathcal{C}_\Omega \\ 0, & \text{otherwise.} \end{cases}$$

Observe that (7.10) now computes $E_{\gamma,\delta,\Omega}$, $\gamma \in \mathcal{D}$, $\Omega \in \mathcal{E}_\delta$. Define $\mathbb{E}_\delta : \mathcal{E}_\delta \to \mathcal{M}_{\mathbf{K}}$ by

$$\mathbb{E}_\delta(\Omega) = \sum_{\gamma \in \mathcal{D}} E_{\gamma,\delta,\Omega}(u)\hat{C}_\gamma.$$

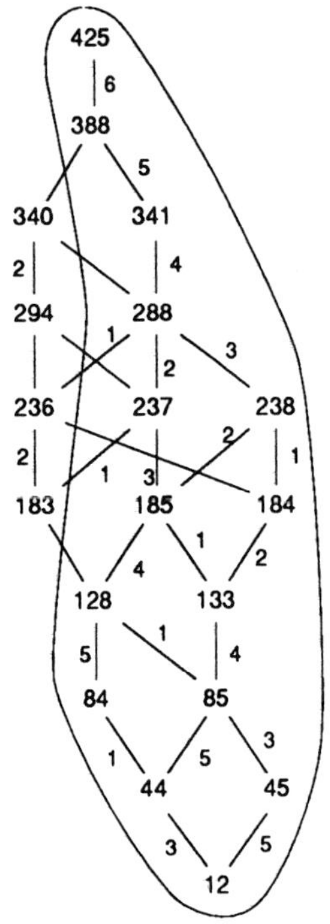

$T_{w_{16^*}}$ and $C_{\emptyset}$

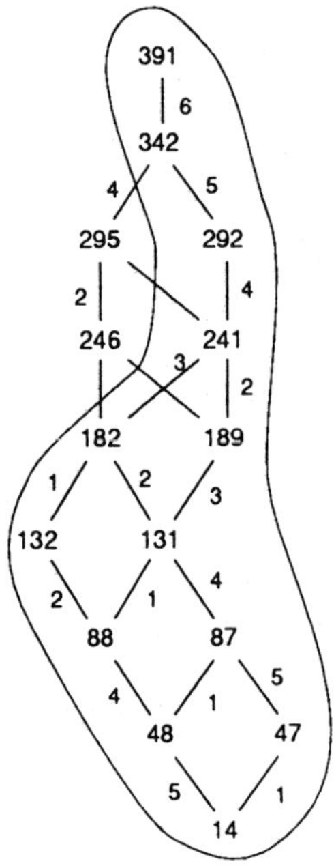

$T_{w_{13^*}}$ and C_{Ω_1}

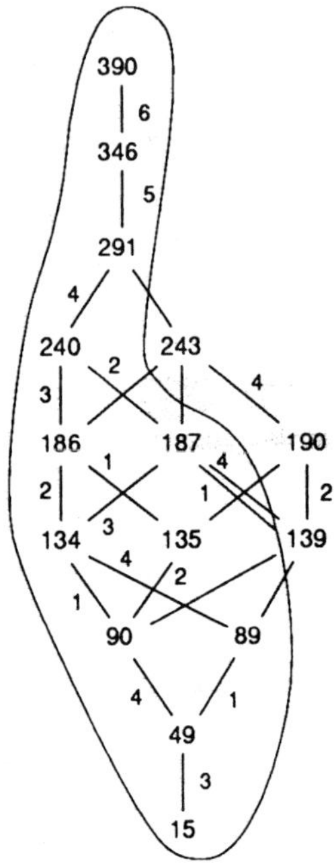

$T_{w_{14^*}}$ and C_{Ω_2}

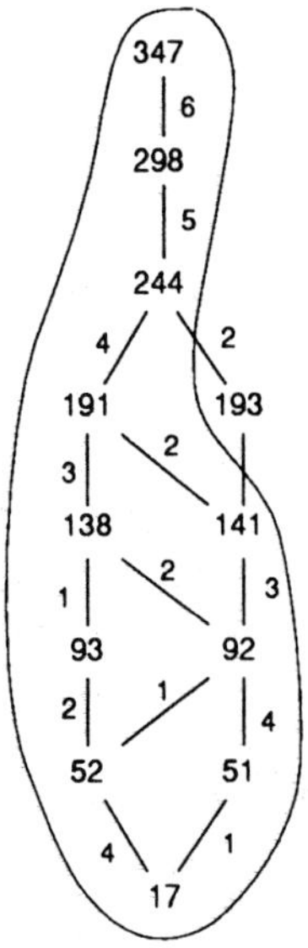

$T_{w_{11^*}}$ and C_{Ω_3}

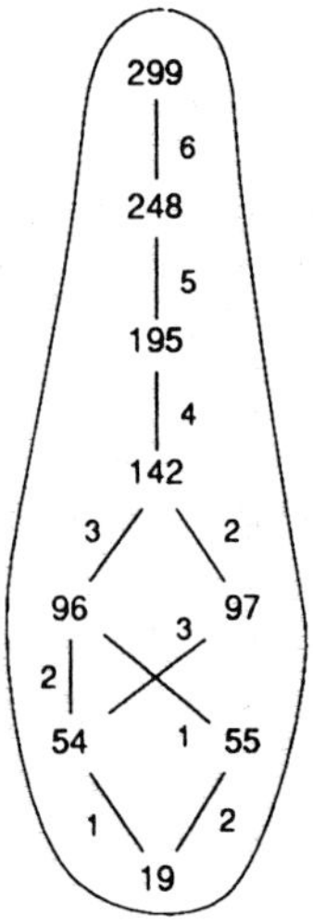

$T_{w_{9^*}}$ and C_{Ω_4}

FIGURE 7.1. Orbit poset examples

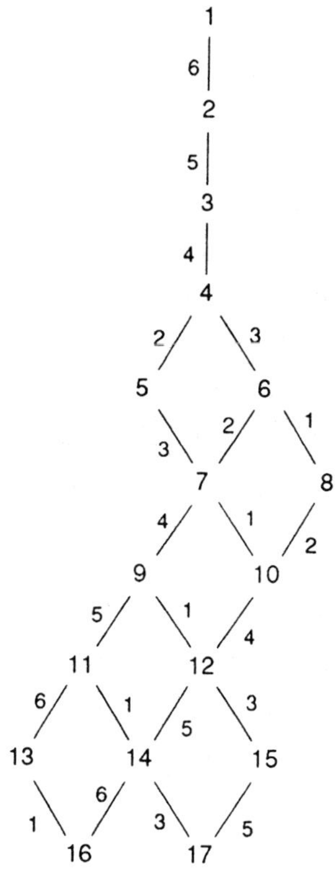

FIGURE 7.2. $\mathcal{C}$ poset

THEOREM 7.11. *Let* $\delta \in \mathcal{P}_{\sigma_i, M}$, $i = 4, 5, 6$ *and put* $\mathcal{E}_\delta = \mathcal{E}_{\mathbb{A}(\delta)}$. *We have*

$$\delta = \sum_{\Omega \in \mathcal{E}_\delta} u^{-|\Omega^+|/2} \mathbb{E}_\delta(\Omega).$$

In summary, the map $\mathbb{E}_\delta$ will attach to each Enright-Shelton family Ω a distinguished collection of $\mathbf{K}$-orbits inside the $\mathbf{Q}^\sharp$-orbit $\mathcal{O}_{\overline{\mathbb{A}(\delta)\tau_\Omega}}$, together with certain "weighting factors" on each orbit. If we are lucky and δ lies in a holomorphic pyramid, then the map $\mathbb{E}_\delta$ is quite simple: it attaches to each Ω one or two such $\mathbf{K}$-orbits, together with weighting factors easily guessed from codimension considerations. On the other hand, as δ moves into more complicated pyramids, many more $\mathbf{K}$-orbits are attached in complicated ways.

To give (7.11) content, consider the following example; recall the parameters introduced in (2.17) and (4.23).

EXAMPLE. The $\mathbf{Q}^\sharp$-orbit $\mathcal{O}_{w_{16^*}}$ contains 19 $\mathbf{K}$-orbits (see Figure 7.1), six of which lie in our pyramids $\mathcal{P}_{\sigma_i, M}$, $i = 4, 5, 6$. We will apply the above algorithm to the parameters:

(7.12) $\qquad \delta_{425} \in \mathcal{P}_{\text{holo}, M}, \quad \delta_{341} \in \mathcal{P}_{\sigma_5, M} \quad \text{and} \quad \delta_{185} \in \mathcal{P}_{\sigma_4, M}.$

To begin with, notice that $\mathbb{A}(\delta_i) = w_{16^*}, i = 425, 341, 185$. We define

$$\mathcal{E}_{\delta_{425}} = \mathcal{E}_{\delta_{341}} = \mathcal{E}_{\delta_{185}} = \mathcal{E}_{w_{16^*}}. \tag{7.13}$$

Then by [11], we have

$$\begin{aligned} \mathcal{E}_{w_{16^*}} &= \{\varnothing, \Omega_1, \Omega_2, \Omega_3, \Omega_4\}; \\ \overline{w_{16^*} r_\varnothing} &= w_{16^*}, |\varnothing| = 0; \\ \overline{w_{16^*} r_{\Omega_1}} &= w_{13^*}, |\Omega_1| = 1; \\ \overline{w_{16^*} r_{\Omega_2}} &= w_{14^*}, |\Omega_2| = 1; \\ \overline{w_{16^*} r_{\Omega_3}} &= w_{11^*}, |\Omega_3| = 2; \\ \overline{w_{16^*} r_{\Omega_4}} &= w_{9^*}, |\Omega_4| = 1. \end{aligned} \tag{7.14}$$

(7.15) *The case of* δ_{425}: Apply the above algorithm and find

$$\begin{aligned} \mathbb{E}_{\delta_{425}}(\varnothing) &= \hat{C}_{\delta_{425}} - u^{-1/2}\hat{C}_{\delta_{388}}; \\ \mathbb{E}_{\delta_{425}}(\Omega_1) &= \hat{C}_{\delta_{391}} - u^{-1/2}\hat{C}_{\delta_{342}}; \\ \mathbb{E}_{\delta_{425}}(\Omega_2) &= \hat{C}_{\delta_{390}} - u^{-1/2}\hat{C}_{\delta_{346}}; \\ \mathbb{E}_{\delta_{425}}(\Omega_3) &= \hat{C}_{\delta_{347}} - u^{-1/2}\hat{C}_{\delta_{298}}; \\ \mathbb{E}_{\delta_{425}}(\Omega_4) &= \hat{C}_{\delta_{299}} - u^{-1/2}\hat{C}_{\delta_{248}}. \end{aligned}$$

hence

$$\begin{aligned} \delta_{425} &= \sum_{\Omega \in \mathcal{E}_{\delta_{425}}} u^{-|\Omega^+|/2}\mathbb{E}_{\delta_{425}}(\Omega) \\ &= \hat{C}_{\delta_{425}} - u^{-1/2}[\hat{C}_{\delta_{391}} + \hat{C}_{\delta_{390}} + \hat{C}_{\delta_{388}} + \hat{C}_{\delta_{299}}] \\ &\quad + u^{-1}[\hat{C}_{\delta_{347}} + \hat{C}_{\delta_{346}} + \hat{C}_{\delta_{342}} + \hat{C}_{\delta_{248}}] - u^{-3/2}\hat{C}_{\delta_{298}}. \end{aligned}$$

(7.16) *The case of* δ_{341}: Apply the above algorithm and find

$$\begin{aligned} \mathbb{E}_{\delta_{341}}(\varnothing) &= \hat{C}_{\delta_{341}} - u^{-1/2}[\hat{C}_{\delta_{288}} + \hat{C}_{\delta_{84}}] + u^{-1}\hat{C}_{\delta_{128}}; \\ \mathbb{E}_{\delta_{341}}(\Omega_1) &= \hat{C}_{\delta_{292}} - u^{-1/2}[\hat{C}_{\delta_{241}} + \hat{C}_{\delta_{47}}] + u^{-1}\hat{C}_{\delta_{87}}; \\ \mathbb{E}_{\delta_{341}}(\Omega_2) &= \hat{C}_{\delta_{291}} - u^{-1/2}\hat{C}_{\delta_{240}} + u^{-1}\hat{C}_{\delta_{89}}; \\ \mathbb{E}_{\delta_{341}}(\Omega_3) &= \hat{C}_{\delta_{244}} - u^{-1/2}\hat{C}_{\delta_{191}} + u^{-1}\hat{C}_{\delta_{51}}; \\ \mathbb{E}_{\delta_{341}}(\Omega_4) &= \hat{C}_{\delta_{195}} - u^{-1/2}\hat{C}_{\delta_{142}}. \end{aligned}$$

hence

$$\begin{aligned} \delta_{341} &= \sum_{\Omega \in \mathcal{E}_{\delta_{341}}} u^{-|\Omega^+|/2}\mathbb{E}_{\delta_{341}}(\Omega) \\ &= \hat{C}_{\delta_{341}} - u^{-1/2}[\hat{C}_{\delta_{292}} + \hat{C}_{\delta_{291}} + \hat{C}_{\delta_{288}} + \hat{C}_{\delta_{195}} + \hat{C}_{\delta_{84}}] \\ &\quad + u^{-1}[\hat{C}_{\delta_{244}} + \hat{C}_{\delta_{241}} + \hat{C}_{\delta_{240}} + \hat{C}_{\delta_{142}} + \hat{C}_{\delta_{128}} + \hat{C}_{\delta_{47}}] \\ &\quad - u^{-3/2}[\hat{C}_{\delta_{191}} + \hat{C}_{\delta_{89}} + \hat{C}_{\delta_{87}}] + u^{-2}\hat{C}_{\delta_{51}}. \end{aligned}$$

(7.17) *The case of* δ_{185}: Apply the above algorithm and find

$$\begin{aligned}
\mathbb{E}_{\delta_{185}}(\varnothing) &= \hat{C}_{\delta_{185}} - u^{-1/2}[\hat{C}_{\delta_{133}} + \hat{C}_{\delta_{128}} + \hat{C}_{\delta_{45}}] + u^{-1}\hat{C}_{\delta_{85}};\\
\mathbb{E}_{\delta_{185}}(\Omega_1) &= \hat{C}_{\delta_{131}} - u^{-1/2}[\hat{C}_{\delta_{88}} + \hat{C}_{\delta_{87}}] + u^{-1}\hat{C}_{\delta_{48}};\\
\mathbb{E}_{\delta_{185}}(\Omega_2) &= \hat{C}_{\delta_{134}} - u^{-1/2}[\hat{C}_{\delta_{90}} + \hat{C}_{\delta_{89}} + \hat{C}_{\delta_{15}}] + u^{-1}\hat{C}_{\delta_{49}};\\
\mathbb{E}_{\delta_{185}}(\Omega_3) &= \hat{C}_{\delta_{92}} - u^{-1/2}[\hat{C}_{\delta_{52}} + \hat{C}_{\delta_{51}}] + u^{-1}\hat{C}_{\delta_{17}};\\
\mathbb{E}_{\delta_{185}}(\Omega_4) &= \hat{C}_{\delta_{54}} - u^{-1/2}\hat{C}_{\delta_{19}}.
\end{aligned}$$

hence

$$\begin{aligned}
\delta_{185} &= \sum_{\Omega\in\mathcal{E}_{\delta_{185}}} u^{-|\Omega^+|/2}\mathbb{E}_{\delta_{185}}(\Omega)\\
&= \hat{C}_{\delta_{185}} - u^{-1/2}[\hat{C}_{\delta_{134}} + \hat{C}_{\delta_{133}} + \hat{C}_{\delta_{131}} + \hat{C}_{\delta_{128}} + \hat{C}_{\delta_{54}} + \hat{C}_{\delta_{45}}]\\
&\quad + u^{-1}[\hat{C}_{\delta_{92}} + \hat{C}_{\delta_{90}} + \hat{C}_{\delta_{89}} + \hat{C}_{\delta_{88}} + \hat{C}_{\delta_{87}} + \hat{C}_{\delta_{85}} + \hat{C}_{\delta_{19}} + \hat{C}_{\delta_{15}}]\\
&\quad - u^{-3/2}[\hat{C}_{\delta_{52}} + \hat{C}_{\delta_{51}} + \hat{C}_{\delta_{49}} + \hat{C}_{\delta_{48}}] + u^{-2}\hat{C}_{\delta_{17}}.
\end{aligned}$$

8. LOEWY LENGTH ESTIMATES

In this chapter, we discuss theoretical and best possible upper bounds for the Loewy length of standard modules attached to parameters in $\mathcal{D}^{\max}$, which, in turn, gives best possible estimates for the Loewy length of generalized principal series in $\mathcal{F}_{\max}$. After discussing the theoretical bounds, we show that whenever $\mathcal{F}_{\max}$ is multiplicity free, then the theoretical bound is best possible if and only if G is quasi-split. We conclude by describing best possible estimates, whenever G is not locally isomorphic to $SO^*(2n)$ or $Sp_n\mathbb{R}, n \geq 4$. This is follwed by a conjecture on best possible bounds, covering all G of Hermitian symmetric type.

If $P = MAN$ is a maximal cuspidal parabolic subgroup and $\mathcal{P}_{\sigma,M}$ any associated pyramid, then (2.4) tells us that $\mathcal{P}_{\sigma,M}$ is in one-to-one correspondence with the "upper half" of the poset $W^{\mathbf{P}}$. This allows us to easily calculate the number

$$t_{\mathbf{M}} = \#\,\text{levels in the poset}\, \mathcal{P}_{\sigma,M};$$

i.e., the ℓ^I length of the unique maximal length element in $\mathcal{P}_{\sigma,M}$. The minimal length elements $\gamma \in \mathcal{P}_{\sigma,M}$ ($\ell^I = 1$) parametrize *Schmid character identities*, as in (2.5), which have Loewy length 2. Arguing as in [15, §4], we see that if a simple reflection s of complex type for $\delta \in \mathcal{D}$ corresponds to a root not in the tau invariant of $\overline{\pi}(\delta)$, then $\ell\ell(\pi(s \circ \delta)) \leq \ell\ell(\pi(\delta)) + 1$. These remarks then imply that the theoretical maximum Loewy length of a standard module attached to a parameter in $\mathcal{P}_{\sigma,M}$ is given by the formula: $t_{\mathbf{M}} + 1 =_{\text{def.}} b_{\mathbf{M}}$. This allows us to define

$$b_{\mathbf{K}} = \max_{\mathbf{M}} b_{\mathbf{M}}, \tag{8.1}$$

where $\mathbf{M}$ ranges over the complexified Levi factors of all maximal cuspidal parabolic subgroups P of G. In summary,

LEMMA 8.2. *The theoretical maximum Loewy length for any standard module attached to a parameter in $\mathcal{D}^{\max}$ is $b_{\mathbf{K}}$. In fact, this bounds the Loewy length of any generalized principal series representation in $\mathcal{F}_{\max}$.*

We have tabulated the values of $b_{\mathbf{K}}$ in Table 8.1. In that same table, we have included the theoretical maximum Loewy length for a generalized Verma module N_w in the category $\mathcal{O}(\mathfrak{g}, \mathfrak{q}^\sharp)$, denoted by $b_{\mathbf{Q}^\sharp}$; observe that $b_{\mathbf{Q}^\sharp} = 1 + \dim_{\mathbb{C}}\mathbf{U}$.

We define

$$\ell\ell_{\mathbf{K},\max} = \max_{\delta \in \mathcal{D}^{\max}} \ell\ell(\pi(\delta)).$$

TABLE 8.1. Loewy length data in $\mathcal{HC}_0$ and $\mathcal{O}(\mathfrak{g}, \mathfrak{q}^{\#})$

Group	Real rank	b_K	$\ell\ell_{K,\max}$	Hermitian pair	$b_{Q^{\#}}$	$\ell\ell_{Q^{\#},\max}$
$SU(p,q)$ $(p \geq q)$	q	$p+q$	$2q, q=p$ $2q+1, p>q$	$(A_{p+q-1}, A_{p-1} \times A_{q-1})$	$pq+1$	$q+1$
$SO_e(2,2n-1)$	2	$2n-1$	3	(B_n, B_{n-1})	$2n$	2
$Sp_n\mathbb{R}$	n	$2n-1$	$(2n-1)^{\natural}$	(C_n, A_{n-1})	$\frac{n(n+1)}{2}+1$	$[\frac{n+1}{2}]+1$
$SO_e(2,2n-2)$	2	$2n-2$	4	(D_n, D_{n-1})	$2n-1$	3
$SO^*(2n)$	$[\frac{n}{2}]$	$2n-2$	$(n)^{\natural}$	(D_n, A_{n-1})	$\frac{n(n-1)}{2}+1$	$[\frac{n}{2}]+1$
$E_{6,-14}$	2	12	5	(E_6, D_5)	17	3
$E_{7,-25}$	3	18	6	(E_7, E_6)	28	4

$^{\natural}$See conjecture I

Obviously, $\ell\ell_{\mathbf{K},\max}$ depends on the group G. Recall that G is said to be *quasi-split* if $M_m A_m$ is a maximally split Cartan subgroup; equivalently, if $\mathbf{P_m}$ is a Borel subgroup of $\mathbf{G}$.

THEOREM G. *Assume G is simple of Hermitian type and $\mathcal{F}_{\max}$ is multiplicity free. Then $\ell\ell_{\mathbf{K},\max} = b_{\mathbf{K}}$ if and only if G is quasi-split.*

We begin by giving the proof, assuming

CLAIM 8.3. *The values of $\ell\ell_{\mathbf{K},\max}$ in Table 8.1 are valid, whenever G is not locally isomorphic to $SO^*(2n)$ or $Sp_n\mathbb{R}$.*

Proof of Theorem G. If G is simple, quasi-split and of Hermitian type, then G is locally isomorphic to one of the following groups:

$$SU(q,q), \quad SU(q+1,q) \quad \text{or} \quad Sp_n\mathbb{R}.$$

Of these cases, via Theorem B, only $SU(q,q), SU(q+1,q), Sp_2\mathbb{R}$ and $Sp_3\mathbb{R}$ satisfy the multiplicity free condition on $\mathcal{F}_{\max}$. If G is not locally isomorphic to $SO^*(2n), Sp_2\mathbb{R}$ or $Sp_3\mathbb{R}$, use Table 8.1 to check that $\ell\ell_{\mathbf{K},\max} = b_{\mathbf{K}}$ if and only if G is quasi-split. By the work in chapter 3, we can see that $\ell\ell_{\mathbf{K},\max} = b_{\mathbf{K}}$ holds for $Sp_2\mathbb{R}$ or $Sp_3\mathbb{R}$. Our proof is complete once we establish:

CLAIM 8.4. *If $G = SO^*(2n)$, then $\ell\ell_{\mathbf{K},\max} < b_{\mathbf{K}}$.*

Recall the parametrization of $\mathcal{D}_o$ for $SO^*(2n)$ in (2.11). By (2.4),we know that $\mathcal{P}_{\sigma,M}$ will be in one-to-one correspondence with $W^{\mathbf{P}}_{upper}$, which is the "upper half" of a $(D_n, A_1 \times D_{n-2})$ poset, as parametrized in [1, Fig. 4.3]. Replacing [1, Fig. 4.3] parameters by those in chapter 2, we recall Figure 4.3, which parametrizes the case when $\mathcal{P}_{\sigma,M} = \mathcal{P}_{\text{holo},M}$. The other pyramids look just like Figure 4.3, but we have to replace the notations $\langle i,j\rangle$ and $((i,j))$ with the new notation

$$\langle i,j\rangle_\sigma \qquad \text{and} \qquad ((i,j))_\sigma,$$

where the subscript σ is keeping track of the assignment of $\pm$-signs attached to the numbers $\{\, k \mid k \neq i, k \neq j \,\}$; this assignment will be determined by the pyramid. (In the case when σ is a holomorphic discrete series, the assignments are those given by (4.22).) Now, choose ω a discrete series on M so that the cardinality of the complement of the tau invariant of ω is as large as possible. Consider the parameter $\langle 1,n\rangle_\omega$ in $\mathcal{P}_{\omega,M}$. Then comparing the description of the $\mathcal{D}_o$ sets for $SU(p,q)$ and $SO^*(2n)$ in (2.7) and (2.11), we check that

(8.5) $\{\delta \preceq \delta_{\langle 1,n\rangle_\omega}\}$ is poset isomorphic to $\mathcal{D}_{SU(n-[(n+1)/2],[(n+1)/2]),o}$.

Now, $\pi(\delta_{\langle 1,n\rangle_\omega})$, viewed as an element of $\mathcal{D}_{SU(n-[(n+1)/2],[(n+1)/2]),o}$ is a pyramid top. Consequently, we can appeal to Table 8.1 in the HS.1 cases to conclude the Loewy length is at most n; in fact, since $SU(n - [(n+1)/2], [(n+1)/2])$ is quasi-split, it is exactly n. Moreover, Theorem E allows us to compute the radical filtration directly and easily to find

(8.6) The bottom layer in the radical filtration of the module $\pi(\delta_{\langle 1,n\rangle_\omega})$ is irreducible with respect to which the simple reflection s_2 is of downward complex type.

The analog of the discussion in [15, §4] for the category $\mathcal{HC}_o$ implies

$$\ell\ell(\pi(\delta_{\langle 1,n\rangle_\omega})) = \ell\ell(\pi(s_2 \circ \delta_{\langle 1,n\rangle_\omega})) = \ell\ell(\pi(\delta_{\langle 2,n\rangle_\omega})) \le n. \tag{8.7}$$

But now, consider any strictly increasing path inside $\mathcal{P}_{\omega,M}$ from $\langle 2,n\rangle_\omega$ to $\delta_{\omega,M}$, which denotes the top of the pyramid $\mathcal{P}_{\omega,M}$; a look at Figure 4.3 shows this path will involve $n-3$ simple reflections of upward complex type. In the worst case senario, we would increase Loewy length exactly by one with each simple wall crossing along this path; again, we are using the analog of the discussion in [15, §4] and the fact that these reflections are of upward complex type. These remarks and (8.7) imply

$$\ell\ell(\pi(\delta_{\omega,M})) \le n + (n-3) = 2n - 3 < 2n - 2.$$

This proves Claim 8.4, when $\sigma = \omega$. For the other pyramids $\mathcal{P}_{\sigma,M}, \sigma \neq \omega$, we can still argue as above, but (8.5) is changed to read

(8.8) The set $\{\delta \preceq \delta_{\langle 1,n\rangle_\sigma}\}$ is poset isomorphic to $\mathcal{D}_{SU(p,q),o}$, for some choice of p and q satisfying $p+q=n$.

We can then combine Theorem E and (8.7) to find that the module $\pi(\delta_{\langle 1,n\rangle_\sigma})$ has Loewy length bounded by that of $\pi(\delta_{\langle 1,n\rangle_\omega})$. This completes the proof of (8.4), modulo the proof of Claim 8.3. □

In the spirit of (8.2), we define

$$\ell\ell_{\mathbf{Q}^\sharp,\max} = \ell\ell(N_{w^{\mathbf{Q}^\sharp}}) \tag{8.9}$$

where $w^{\mathbf{Q}^\sharp}$ is the longest element of $W^{\mathbf{Q}^\sharp}$; then $\ell\ell_{\mathbf{Q}^\sharp,\max}$ is the longest Loewy length achieved by a generalized Verma module in the category $\mathcal{O}(\mathfrak{g}, \mathfrak{q}^\sharp)$. The values of $\ell\ell_{\mathbf{Q}^\sharp,\max}$ are given in Table 8.1 and follow from [10, Table 2.1]. Comparing with the values of $b_{\mathbf{Q}^\sharp}$, we see that the theoretical and best possible least upper bounds on Loewy length coincide if and only if $G = SU(1,1)$; in this sense, Theorem G has no analog in the category $\mathcal{O}(\mathfrak{g}, \mathfrak{q}^\sharp)$ setting. However, notice that $\ell\ell_{\mathbf{Q}^\sharp,\max}$ is dominated by a linear function in the real rank. The analog of this fact does hold in the Harish-Chandra setting and is spelled out in the next result.

THEOREM H. *Let G be simple of Hermitian type. Then there exists a linear function $p_G(r)$, in the real rank r of G, so that $\ell\ell(\pi) \leq p_G(r)$, for all $\pi \in \mathcal{F}_{\max}$.*

If G=$Sp_n\mathbb{R}$, we conjecture that $p_G(n) = 2n - 1 = \ell\ell_{\mathbf{K},\max} = b_{\mathbf{K}}$. Note that this is the case for $n = 2, 3$, by our discussions in chapter 3. Computer calculations have verified this is the case for $n \leq 5$. If $G = SO^*(2n)$, we conjecture $\ell\ell_{\mathbf{K},\max} = n$. This has also been verified using the computer, when $n \leq 7$. We formalize this as

CONJECTURE I. *The values for $\ell\ell_{\mathbf{K},\max}$ are as given in Table 8.1.*

Proof of Claim 8.3. Assume G is not locally isomorphic to $SO^*(2n)$ or $Sp_n\mathbb{R}$. In the cases of $SO_e(2,k)$ we appeal to Theorem 4.8 and Lemma 5.1. The case of $E_{6,-14}$ is discussed in chapter 7 and $E_{7,-25}$ is handled in the same way, via computer.

This leaves us with the case of $SU(p,q)$ and we assume G is of this type for the remainder of the section. The key point is that we can fairly easily describe the standard composition factors of the standard modules attached to the tops of the various pyramids; i.e., the modules $\pi(\delta_{\sigma,M})$). If σ is a discrete series for the Levi factor M of a maximal cuspidal parabolic subgroup P of G, then define

$$F_\sigma = \{\gamma \in \hat{\mathcal{P}}_{\sigma,M} \mid \ell(\gamma) = \ell(\delta_{\sigma,M}) - 1\}$$

and let $m_\sigma = |F_\sigma|$.

CLAIM 8.10. *(i) There exists a unique set $\Omega_o \subset \mathcal{E}_{\delta_{\sigma,M}}$ such that $|\Omega_o^+| = m_\sigma$ and $|\Omega^+| < m_\sigma$ for all other $\Omega \subset \mathcal{E}_{\delta_{\sigma,M}}$.*

(ii) $\ell\ell(\pi(\delta_{\sigma,M})) = m_\sigma + 1$.

(iii) If G=SU(q,q), then $m_\sigma \leq 2q - 1$ and there exists a σ' such that $m_{\sigma'} = 2q - 1$.

(iv) If G=SU(q+1,q), then $m_\sigma \leq 2q$ and there exists a σ' such that $m_{\sigma'} = 2q$.

Part (ii) of (8.10) is a direct consequence of Theorem E and (8.10)(i). For part (i), we recall that the non-weak order relations down from the top $\delta_{\sigma,M}$ inside the G-pyramid $\mathcal{P}_{\sigma,M}$ have the form

$$
(8.11)\qquad
\begin{array}{ccc}
(1,n)\cdots i^{+}(i+1)^{-}\cdots & & (1,n)\cdots i^{-}(i+1)^{+}\cdots \\
{}_{r_i}\searrow & & \swarrow_{r_i} \\
& (1,i)(i+1,n)\ \cdots &
\end{array}
$$

where r_i is the fake reflection in Lemma 6.1 relating $\delta_{\sigma,M}$ and $\delta_{\sigma,M}\star r_i$. Then we can describe the elements in F_σ as

$$F_\sigma = \{\, s_1 \times \delta_{\sigma,M},\ s_{n-1} \times \delta_{\sigma,M} \} \cup \{\, \delta_{\sigma,M} \star r_i \mid r_i \text{ is as in } (8.11)\, \}.$$

Let $S_\sigma = \{\, r_i \text{ in } (8.11)\, \} \cup \{s_1, s_{n-1}\}$. Every $R \subset S_\sigma$ determines a set Ω_R allowed in Theorem E. We know that this accounts for all the standard composition factors of $\pi(\delta_{\sigma,M})$ and there will be a unique standard factor in layer m_σ of the radical filtration, corresponding to the case Ω_{S_σ}. But the conditions on a set Ω in chapter 6 and the form of the tuple $\delta_{\sigma,M}$ insure that Ω_{S_σ} is the unique $\Omega \in \mathcal{E}_{\delta_{\sigma,M}}$ such that $|\Omega^+| = m_\sigma$. This proves (8.10)(i).

From our work in chapter 6, it is clear that

$$m_\sigma = 2 + \text{the cardinality of the complement of the tau invariant of } \sigma.$$

The number 2 is coming from the two complex downward type reflections attached to $\delta_{\sigma,M}$ in $\mathcal{P}_{\sigma,M}$. In turn, $m_\sigma = 2 + l$, where $l =$ number of non-compact simple reflections in the set of simple roots determined by the M discrete series representation σ. We know that each σ corresponds to a parameter in the poset $W_{\mathbf{P}}^{\mathbf{K_M}}$; this is just the Hermitian symmetric diagram attached to $SU(p-1,q-1)$. The diagram descriptions given in [8] allow one to compute m_σ easily and arrive at

$$(8.12)\qquad \begin{array}{c} \text{If } M_{ss} = SU(q-1,q-1) \text{ (resp. } SU(p-1,q-1) \text{ for } p > q), \\ \text{then } l \le 2q-3 \text{ (resp. } 2q-2). \end{array}$$

By (8.12) we arrive at the upper bounds in (iii) and (iv). In addition, it is important to note that there exists at least one (possibly several) $\sigma' \in W_{\mathbf{P}}^{\mathbf{K_M}}$ where the upper bounds in (iii) and (iv) are realized. This completes the proof of (8.10) and establishes (8.3). □

9. APPENDIX: EXCEPTIONAL DATA

In this chapter, we give computer-generated descriptions of the weak $\preceq_w$ order on $\mathcal{D}$ for $G = E_{6,-14}$ and $E_{7,-25}$. The data is presented according to the conventions below.

(9.1a) In the "Parameter column" of Table 9.1 (resp. 9.2), we enumerate the set $\mathcal{D}$ via numbers "i", $1 \le i \le 513$ (resp. 3017);

(b) In the "ℓ^I column", we give $\ell^I(\delta_i)$; ("ℓ^I" is abbreviated "ℓ" in Table 9.2, due to space limitations);

(c) The "dim A column" determines the Cartan subgroup, via Table 2.3 (resp. 2.4); ("dim A" is abbreviated simply "d" in Table 9.2).

The "Simple columns" $1, 2, \ldots, 6$ (resp. 7) and the "Other column" are describing the $\mathcal{H}$ structure on $\mathcal{M}_{\mathbf{K}}$. To interpret this, fix a row "i" of the table, corresponding to δ_i. We now describe what the entries in the columns of this row mean:

(d) If "-1"appears in the jth-simple column, then s_j is of compact type for δ_i.

(e) Assume the entry in the jth-simple column is $k \neq -1$ and there is an entry $r(j)$ in the "Other column" of row i. Then s_j is of real type I for δ_i, $s_j \circ \delta_k = s_j \circ \delta_r = \delta_i$, and $s_j \times \delta_k = \delta_r$.

(f) Assume the entry in the jth-simple column is $k \neq -1$ and there is an entry $r(j)$ in the "Other column" **of row k**. Then s_j is of noncompact type I for δ_i, and $s_j \circ \delta_i = \delta_k$.

(g) Assume the entry in the jth-simple column is $k \neq -1$ and neither row i nor row k contains an entry $r(j)$ in the "Other column". Then s_j is of complex type for δ_i and $s_j \times \delta_i = \delta_k$.

Taken collectively, the rules in (9.1) determine the weak $\preceq_w$ order on $\mathcal{D}$.

TABLE 2.5. $\mathcal{D}$ diagram for $E_{6,-14}$

Parameter	ℓ^I	dim A	Simples 1	2	3	4	5	6	Other
1	0	0	28	-1	-1	-1	-1	-1	
2	0	0	28	-1	29	-1	-1	-1	
3	0	0	-1	-1	29	30	-1	-1	
4	0	0	-1	31	-1	30	32	-1	
5	0	0	-1	31	-1	-1	33	-1	
6	0	0	-1	34	-1	-1	32	35	
7	0	0	-1	34	-1	36	33	37	
8	0	0	-1	38	-1	-1	-1	35	
9	0	0	-1	-1	39	36	-1	40	
10	0	0	-1	38	-1	41	-1	37	
11	0	0	42	-1	39	-1	-1	43	
12	0	0	-1	-1	44	41	45	40	
13	0	0	42	-1	-1	-1	-1	46	
14	0	0	47	-1	44	-1	48	43	
15	0	0	-1	-1	49	-1	45	-1	
16	0	0	47	-1	-1	-1	50	46	
17	0	0	51	-1	49	52	48	-1	
18	0	0	51	-1	-1	53	50	-1	
19	0	0	54	55	-1	52	-1	-1	
20	0	0	54	56	57	53	-1	-1	
21	0	0	58	55	-1	-1	-1	-1	
22	0	0	58	56	59	-1	-1	-1	
23	0	0	-1	60	57	-1	-1	-1	
24	0	0	-1	60	59	61	-1	-1	
25	0	0	-1	-1	-1	61	62	-1	
26	0	0	-1	-1	-1	-1	62	63	
27	0	0	-1	-1	-1	-1	-1	63	
28	1	1	2	-1	64	-1	-1	-1	1(1)
29	1	1	64	-1	3	65	-1	-1	2(3)
30	1	1	-1	66	65	4	67	-1	3(4)
31	1	1	-1	5	-1	66	68	-1	4(2)
32	1	1	-1	68	-1	67	6	69	4(5)
33	1	1	-1	68	-1	70	7	71	5(5)
34	1	1	-1	7	-1	72	68	73	6(2)
35	1	1	-1	73	-1	-1	69	8	6(6)
36	1	1	-1	72	74	9	70	75	7(4)
37	1	1	-1	73	-1	75	71	10	7(6)
38	1	1	-1	10	-1	76	-1	73	8(2)
39	1	1	77	-1	11	74	-1	78	9(3)
40	1	1	-1	-1	78	75	79	12	9(6)
41	1	1	-1	76	80	12	81	75	10(4)
42	1	1	13	-1	77	-1	-1	82	11(1)
43	1	1	82	-1	78	-1	83	14	11(6)
44	1	1	84	-1	14	80	85	78	12(3)
45	1	1	-1	-1	85	81	15	79	12(5)
46	1	1	82	-1	-1	-1	86	16	13(6)
47	1	1	16	-1	84	-1	87	82	14(1)
48	1	1	87	-1	85	88	17	83	14(5)
49	1	1	89	-1	17	90	85	-1	15(3)
50	1	1	87	-1	-1	91	18	86	16(5)
51	1	1	18	-1	89	92	87	-1	17(1)
52	1	1	92	93	90	19	88	-1	17(4)
53	1	1	92	94	95	20	91	-1	18(4)
54	1	1	20	96	97	92	-1	-1	19(1)
55	1	1	96	21	-1	93	-1	-1	19(2)
56	1	1	96	22	98	94	-1	-1	20(2)
57	1	1	97	98	23	95	-1	-1	20(3)
58	1	1	22	96	99	-1	-1	-1	21(1)
59	1	1	99	98	24	100	-1	-1	22(3)
60	1	1	-1	24	98	101	-1	-1	23(2)
61	1	1	-1	101	100	25	102	-1	24(4)
62	1	1	-1	-1	-1	102	26	103	25(5)
63	1	1	-1	-1	-1	-1	103	27	26(6)

Parameter	ℓ^I	dim A	Simples 1	2	3	4	5	6	Other
64	2	1	29	-1	28	104	-1	-1	
65	2	1	104	105	30	29	106	-1	
66	2	1	-1	30	105	31	107	-1	
67	2	1	-1	107	106	32	30	108	
68	2	2	-1	33	-1	109	34	110	32(2) 31(5)
69	2	1	-1	110	-1	108	35	32	
70	2	1	-1	111	112	33	36	113	
71	2	1	-1	110	-1	113	37	33	
72	2	1	-1	36	114	34	111	115	
73	2	2	-1	37	1	115	110	38	35(2) 34(6)
74	2	1	116	114	36	39	112	117	
75	2	2	-1	115	117	40	118	41	37(4) 36(6)
76	2	1	-1	41	119	38	120	115	
77	2	1	39	-1	42	116	-1	121	
78	2	2	121	-1	43	117	122	44	40(3) 39(6)
79	2	1	-1	-1	122	123	40	45	
80	2	1	124	119	41	44	125	117	
81	2	1	-1	120	125	45	41	123	
82	2	2	46	-1	121	-1	126	47	43(1) 42(6)
83	2	1	126	-1	122	127	43	48	
84	2	1	44	-1	47	124	128	121	
85	2	2	128	-1	48	129	49	122	45(3) 44(5)
86	2	1	126	-1	-1	130	46	50	
87	2	2	50	-1	128	131	51	126	48(1) 47(5)
88	2	1	131	132	133	48	52	127	
89	2	1	49	-1	51	134	128	-1	
90	2	1	134	135	52	49	133	-1	
91	2	1	131	136	137	50	53	130	
92	2	2	53	138	139	54	131	-1	52(1) 51(4)
93	2	1	138	52	135	55	132	-1	
94	2	1	138	53	140	56	136	-1	
95	2	1	141	140	53	57	137	-1	
96	2	2	56	58	142	138	-1	-1	55(1) 54(2)
97	2	1	57	142	54	141	-1	-1	
98	2	2	142	59	60	143	-1	-1	57(2) 56(3)
99	2	1	59	142	58	144	-1	-1	
100	2	1	144	145	61	59	146	-1	
101	2	1	-1	61	145	60	147	-1	
102	2	1	-1	147	146	62	61	148	
103	2	1	-1	-1	-1	148	63	62	
104	3	1	65	149	-1	64	150	-1	
105	3	1	149	65	66	-1	151	-1	
106	3	1	150	151	67	-1	65	152	
107	3	1	-1	67	151	153	66	154	
108	3	1	-1	154	152	69	-1	67	
109	3	2	-1	153	155	68	153	156	
110	3	2	-1	71	-1	156	73	68	69(2)
111	3	1	-1	70	157	153	72	158	
112	3	1	159	157	70	-1	74	160	
113	3	1	-1	158	160	71	161	70	
114	3	1	162	74	72	-1	157	163	
115	3	2	-1	75	163	73	164	76	72(6)
116	3	1	74	162	-1	77	159	165	
117	3	2	165	163	75	78	166	80	74(6)
118	3	2	-1	164	166	161	75	161	
119	3	1	167	80	76	-1	168	163	
120	3	1	-1	81	168	-1	76	169	
121	3	2	78	-1	82	165	170	84	77(6)
122	3	2	170	-1	83	171	78	85	79(3)
123	3	1	-1	169	172	79	161	81	
124	3	1	80	167	-1	84	173	165	
125	3	1	173	168	81	174	80	172	
126	3	2	86	-1	170	175	82	87	83(1)

Parameter	ℓ^I	dim A	Simples 1	2	3	4	5	6	Other
127	3	1	175	176	177	83	-1	88	
128	3	2	85	-1	87	178	89	170	84(5)
129	3	2	178	179	174	85	174	171	
130	3	1	175	180	181	86	-1	91	
131	3	2	91	182	183	87	92	175	88(1)
132	3	1	182	88	184	-1	93	176	
133	3	1	185	184	88	174	90	177	
134	3	1	90	186	187	89	185	-1	
135	3	1	186	90	93	-1	184	-1	
136	3	1	182	91	188	-1	94	180	
137	3	1	189	188	91	-1	95	181	
138	3	2	94	92	190	96	182	-1	93(1)
139	3	2	187	190	92	187	183	-1	
140	3	1	191	95	94	192	188	-1	
141	3	1	95	191	187	97	189	-1	
142	3	2	98	99	96	193	-1	-1	97(2)
143	3	2	193	192	192	98	194	-1	
144	3	1	100	195	-1	99	196	-1	
145	3	1	195	100	101	192	197	-1	
146	3	1	196	197	102	-1	100	198	
147	3	1	-1	102	197	-1	101	199	
148	3	1	-1	199	198	103	-1	102	
149	4	1	105	104	-1	-1	200	-1	
150	4	1	106	200	-1	-1	104	201	
151	4	1	200	106	107	202	105	203	
152	4	1	201	203	108	-1	-1	106	
153	4	2	-1	109	204	111	109	205	107(4)
154	4	1	-1	108	203	205	-1	107	
155	4	2	206	204	109	-1	204	207	
156	4	2	-1	205	207	110	208	109	
157	4	1	209	112	111	210	114	211	
158	4	1	-1	113	211	205	212	111	
159	4	1	112	209	-1	-1	116	213	
160	4	1	213	211	113	-1	214	112	
161	4	2	-1	212	214	118	123	118	113(5)
162	4	1	114	116	-1	-1	209	215	
163	4	2	215	117	115	-1	216	119	114(6)
164	4	2	-1	118	216	208	115	212	
165	4	2	117	215	-1	121	217	124	116(6)
166	4	2	217	216	118	218	117	214	
167	4	1	119	124	-1	-1	219	215	
168	4	1	219	125	120	220	119	221	
169	4	1	-1	123	221	-1	212	120	
170	4	2	122	-1	126	222	121	128	
171	4	2	222	223	224	122	218	129	
172	4	1	225	221	123	224	214	125	
173	4	1	125	219	-1	226	124	225	
174	4	2	226	227	129	133	129	224	125(4)
175	4	2	130	228	229	126	-1	131	127(1)
176	4	1	228	127	230	-1	-1	132	
177	4	1	231	230	127	224	-1	133	
178	4	2	129	232	233	128	226	222	
179	4	2	232	129	227	-1	227	223	
180	4	1	228	130	234	-1	-1	136	
181	4	1	235	234	130	-1	-1	137	
182	4	2	136	131	236	-1	138	228	132(1)
183	4	2	237	236	131	233	139	229	
184	4	1	238	133	132	239	135	230	
185	4	1	133	238	237	226	134	231	
186	4	1	135	134	240	-1	238	-1	
187	4	2	139	240	141	139	237	-1	134(3)
188	4	1	241	137	136	242	140	234	
189	4	1	137	241	237	-1	141	235	
190	4	2	240	139	138	243	236	-1	
191	4	1	140	141	240	244	241	-1	
192	4	2	244	143	143	145	245	-1	140(4)

Parameter	ℓ^I	dim A	Simples 1	2	3	4	5	6	Other
193	4	2	143	244	243	142	246	-1	
194	4	2	246	245	245	-1	143	247	
195	4	1	145	144	-1	244	248	-1	
196	4	1	146	248	-1	-1	144	249	
197	4	1	248	146	147	250	145	251	
198	4	1	249	251	148	-1	-1	146	
199	4	1	-1	148	251	-1	-1	147	
200	5	1	151	150	-1	252	149	253	
201	5	1	152	253	-1	-1	-1	150	
202	5	1	252	-1	254	151	-1	255	
203	5	1	253	152	154	255	-1	151	
204	5	2	256	155	153	254	155	257	
205	5	2	-1	156	257	158	258	153	154(4)
206	5	2	155	256	-1	-1	256	259	
207	5	2	259	257	156	-1	260	155	
208	5	2	-1	258	260	164	156	258	
209	5	1	157	159	-1	261	162	262	
210	5	1	261	-1	254	157	-1	263	
211	5	1	262	160	158	263	264	157	
212	5	2	-1	161	264	258	169	164	158(5)
213	5	1	160	262	-1	-1	265	159	
214	5	2	265	264	161	266	172	166	160(5)
215	5	2	163	165	-1	-1	267	167	162(6)
216	5	2	267	166	164	268	163	264	
217	5	2	166	267	-1	269	165	265	
218	5	2	269	270	266	166	171	266	
219	5	1	168	173	-1	271	167	272	
220	5	1	271	273	-1	168	-1	274	
221	5	1	272	172	169	274	264	168	
222	5	2	171	275	276	170	269	178	
223	5	2	275	171	277	-1	270	179	
224	5	2	278	277	171	177	266	174	172(4)
225	5	1	172	272	-1	278	265	173	
226	5	2	174	279	280	185	178	278	173(4)
227	5	2	279	174	179	273	179	277	
228	5	2	180	175	281	-1	-1	182	176(1)
229	5	2	282	281	175	276	-1	183	
230	5	1	283	177	176	284	-1	184	
231	5	1	177	283	282	278	-1	185	
232	5	2	179	178	285	-1	279	275	
233	5	2	280	285	178	183	280	276	
234	5	1	286	181	180	287	-1	188	
235	5	1	181	286	282	-1	-1	189	
236	5	2	288	183	182	289	190	281	
237	5	2	183	288	189	280	187	282	185(3)
238	5	1	184	185	288	290	186	283	
239	5	1	290	273	-1	184	-1	284	
240	5	2	190	187	191	291	288	-1	186(3)
241	5	1	188	189	288	292	191	286	
242	5	1	292	-1	-1	188	293	287	
243	5	2	291	291	193	190	294	-1	
244	5	2	192	193	291	195	295	-1	191(4)
245	5	2	295	194	194	293	192	296	
246	5	2	194	295	294	-1	193	297	
247	5	2	297	296	296	-1	-1	194	
248	5	1	197	196	-1	298	195	299	
249	5	1	198	299	-1	-1	-1	196	
250	5	1	298	-1	-1	197	293	300	
251	5	1	299	198	199	300	-1	197	
252	6	1	202	-1	301	200	-1	302	
253	6	1	203	201	-1	302	-1	200	
254	6	2	303	-1	210	204	-1	304	202(3)
255	6	1	302	-1	304	203	305	202	
256	6	2	204	206	-1	303	206	306	
257	6	2	306	207	205	304	307	204	
258	6	2	-1	208	307	212	205	208	

Parameter	ℓ^I	dim A	Simples 1	2	3	4	5	6	Other
259	6	2	207	306	-1	-1	308	206	
260	6	2	308	307	208	309	207	307	
261	6	1	210	-1	310	209	-1	311	
262	6	1	211	213	-1	311	312	209	
263	6	1	311	-1	304	211	313	210	
264	6	2	312	214	212	314	221	216	211(5)
265	6	2	214	312	-1	315	225	217	213(5)
266	6	2	315	316	218	214	224	218	
267	6	2	216	217	-1	317	215	312	
268	6	2	317	318	309	216	-1	314	
269	6	2	218	319	320	217	222	315	
270	6	2	319	218	316	318	223	316	
271	6	1	220	321	322	219	-1	323	
272	6	1	221	225	-1	323	312	219	
273	6	2	321	239	-1	227	-1	324	220(2)
274	6	1	323	324	-1	221	325	220	
275	6	2	223	222	326	-1	319	232	
276	6	2	327	326	222	229	320	233	
277	6	2	328	224	223	324	316	227	
278	6	2	224	328	327	231	315	226	225(4)
279	6	2	227	226	329	321	232	328	
280	6	2	233	329	226	237	233	327	
281	6	2	330	229	228	331	-1	236	
282	6	2	229	330	235	327	-1	237	231(3)
283	6	1	230	231	330	332	-1	238	
284	6	1	332	324	-1	230	333	239	
285	6	2	329	233	232	334	329	326	
286	6	1	234	235	330	335	-1	241	
287	6	1	335	-1	-1	234	336	242	
288	6	2	236	237	241	337	240	330	238(3)
289	6	2	337	334	-1	236	338	331	
290	6	1	239	321	339	238	-1	332	
291	6	2	243	243	244	240	340	-1	
292	6	1	242	-1	341	241	342	335	
293	6	2	342	-1	-1	245	250	343	242(5)
294	6	2	340	340	246	338	243	344	
295	6	2	245	246	340	342	244	345	
296	6	2	345	247	247	343	-1	245	
297	6	2	247	345	344	-1	-1	246	
298	6	1	250	-1	346	248	342	347	
299	6	1	251	249	-1	347	-1	248	
300	6	1	347	-1	-1	251	348	250	
301	7	1	349	-1	252	-1	-1	350	
302	7	1	255	-1	350	253	351	252	
303	7	2	254	-1	349	256	-1	352	
304	7	2	352	-1	263	257	353	254	255(3)
305	7	1	351	-1	353	-1	255	-1	
306	7	2	257	259	-1	352	354	256	
307	7	2	354	260	258	355	257	260	
308	7	2	260	354	-1	356	259	354	
309	7	2	356	357	268	260	-1	355	
310	7	1	349	-1	261	-1	-1	358	
311	7	1	263	-1	358	262	359	261	
312	7	2	264	265	-1	360	272	267	262(5)
313	7	1	359	-1	353	361	263	-1	
314	7	2	360	362	355	264	361	268	
315	7	2	266	363	364	265	278	269	
316	7	2	363	266	270	362	277	270	
317	7	2	268	365	366	267	-1	360	
318	7	2	365	268	357	270	-1	362	
319	7	2	270	269	367	365	275	363	
320	7	2	364	367	269	-1	276	364	
321	7	2	273	290	368	279	-1	369	271(2)
322	7	1	-1	368	271	-1	-1	370	
323	7	1	274	369	370	272	371	271	
324	7	2	369	284	-1	277	372	273	274(2)
325	7	1	371	372	-1	361	274	-1	
326	7	2	373	276	275	374	367	285	
327	7	2	276	373	278	282	364	280	
328	7	2	277	278	373	369	363	279	
329	7	2	285	280	279	375	285	373	
330	7	2	281	282	286	376	-1	288	283(3)
331	7	2	376	374	-1	281	377	289	
332	7	1	284	369	378	283	379	290	
333	7	1	379	372	-1	-1	284	-1	
334	7	2	375	289	-1	285	380	374	
335	7	1	287	-1	381	286	382	292	
336	7	1	382	-1	-1	-1	287	383	
337	7	2	289	375	384	288	385	376	
338	7	2	385	380	-1	294	289	386	
339	7	1	-1	368	290	384	-1	378	
340	7	2	294	294	295	385	291	387	
341	7	1	-1	-1	292	384	388	381	
342	7	2	293	-1	388	295	298	389	292(5)
343	7	2	389	-1	-1	296	383	293	
344	7	2	387	387	297	386	-1	294	
345	7	2	296	297	387	389	-1	295	
346	7	1	-1	-1	298	-1	388	390	
347	7	1	300	-1	390	299	391	298	
348	7	1	391	-1	-1	-1	300	383	
349	8	2	310	-1	303	-1	-1	392	301(1)
350	8	1	392	-1	302	-1	393	301	
351	8	1	305	-1	393	-1	302	-1	
352	8	2	304	-1	392	306	394	303	
353	8	2	394	-1	313	395	304	-1	305(3)
354	8	2	307	308	-1	396	306	308	
355	8	2	396	397	314	307	395	309	
356	8	2	309	398	399	308	-1	396	
357	8	2	398	309	318	397	-1	397	
358	8	1	392	-1	311	-1	400	310	
359	8	1	313	-1	400	401	311	-1	
360	8	2	314	402	403	312	401	317	
361	8	2	401	404	395	325	314	-1	313(4)
362	8	2	402	314	397	316	404	318	
363	8	2	316	315	405	402	328	319	
364	8	2	320	405	315	-1	327	320	
365	8	2	318	317	406	319	-1	402	
366	8	2	399	406	317	-1	-1	403	
367	8	2	405	320	319	407	326	405	
368	8	2	-1	339	321	408	-1	409	322(2)
369	8	2	324	332	409	328	410	321	323(2)
370	8	1	-1	409	323	-1	411	322	
371	8	1	325	410	411	401	323	-1	
372	8	2	410	333	-1	404	324	-1	325(2)
373	8	2	326	327	328	412	405	329	
374	8	2	412	331	-1	326	413	334	
375	8	2	334	337	408	329	414	412	
376	8	2	331	412	415	330	416	337	
377	8	2	416	413	-1	-1	331	417	
378	8	1	-1	409	332	415	418	339	
379	8	1	333	410	418	-1	332	-1	
380	8	2	414	338	-1	414	334	419	
381	8	1	-1	-1	335	415	420	341	
382	8	1	336	-1	420	-1	335	421	
383	8	2	421	-1	-1	-1	343	348	336(6)
384	8	2	-1	408	337	341	422	415	339(4)
385	8	2	338	414	422	340	337	423	
386	8	2	423	419	-1	344	417	338	
387	8	2	344	344	345	423	-1	340	
388	8	2	-1	-1	342	422	346	424	341(5)
389	8	2	343	-1	424	345	421	342	
390	8	1	-1	-1	347	-1	425	346	

Parameter	ℓ^I	dim A	Simples 1	2	3	4	5	6	Other
391	8	1	348	-1	425	-1	347	421	
392	9	2	358	-1	352	-1	426	349	350(1)
393	9	1	426	-1	351	427	350	-1	
394	9	2	353	-1	426	428	352	-1	
395	9	2	428	429	361	353	355	-1	
396	9	2	355	430	431	354	428	356	
397	9	2	430	355	362	357	429	357	
398	9	2	357	356	432	430	-1	430	
399	9	2	366	432	356	-1	-1	431	
400	9	1	426	-1	359	433	358	-1	
401	9	2	361	434	435	371	360	-1	359(4)
402	9	2	362	360	436	363	434	365	
403	9	2	431	436	360	-1	435	366	
404	9	2	434	361	429	372	362	-1	
405	9	2	367	364	363	437	373	367	
406	9	2	432	366	365	438	-1	436	
407	9	2	437	-1	438	367	439	437	
408	9	2	-1	384	375	368	440	441	
409	9	2	-1	378	369	441	442	368	370(2)
410	9	2	372	379	442	434	369	-1	371(2)
411	9	1	-1	442	371	443	370	-1	
412	9	2	374	376	441	373	444	375	
413	9	2	444	377	-1	439	374	445	
414	9	2	380	385	440	380	375	446	
415	9	2	-1	441	376	381	447	384	378(4)
416	9	2	377	444	447	-1	376	448	
417	9	2	448	445	-1	-1	386	377	
418	9	1	-1	442	379	449	378	-1	
419	9	2	446	386	-1	446	445	380	
420	9	1	-1	-1	382	450	381	451	
421	9	2	383	-1	451	-1	389	391	382(6)
422	9	2	-1	440	385	388	384	452	
423	9	2	386	446	452	387	448	385	
424	9	2	-1	-1	389	452	451	388	
425	9	1	-1	-1	391	453	390	451	
426	10	2	400	-1	394	454	392	-1	393(1)
427	10	1	454	455	-1	393	-1	-1	
428	10	2	395	456	457	394	396	-1	
429	10	2	456	395	404	-1	397	-1	
430	10	2	397	396	458	398	456	398	
431	10	2	403	458	396	-1	457	399	
432	10	2	406	399	398	459	-1	458	
433	10	1	454	460	461	400	-1	-1	
434	10	2	404	401	462	410	402	-1	
435	10	2	457	462	401	461	403	-1	
436	10	2	458	403	402	463	462	406	
437	10	2	407	-1	463	405	464	407	
438	10	2	459	-1	407	406	465	463	
439	10	2	464	-1	465	413	407	466	
440	10	2	-1	422	414	-1	408	467	
441	10	2	-1	415	412	409	468	408	
442	10	2	-1	418	410	469	409	-1	411(2)
443	10	1	-1	470	461	411	-1	-1	
444	10	2	413	416	468	464	412	471	
445	10	2	471	417	-1	466	419	413	
446	10	2	419	423	467	419	471	414	
447	10	2	-1	468	416	472	415	473	
448	10	2	417	471	473	-1	423	416	
449	10	1	-1	474	-1	418	472	-1	
450	10	1	-1	475	-1	420	472	476	
451	10	2	-1	-1	421	476	424	425	420(6)
452	10	2	-1	467	423	424	473	422	
453	10	1	-1	477	-1	425	-1	476	
454	11	2	433	478	479	426	-1	-1	427(1)
455	11	1	478	427	-1	-1	-1	-1	
456	11	2	429	428	480	-1	430	-1	
457	11	2	435	480	428	479	431	-1	
458	11	2	436	431	430	481	480	432	
459	11	2	438	-1	481	432	482	481	
460	11	1	478	433	483	-1	-1	-1	
461	11	2	479	483	443	435	-1	-1	433(3)
462	11	2	480	435	434	484	436	-1	
463	11	2	481	-1	437	436	485	438	
464	11	2	439	-1	485	444	437	486	
465	11	2	482	-1	439	-1	438	487	
466	11	2	486	-1	487	445	486	439	
467	11	2	-1	452	446	-1	488	440	
468	11	2	-1	447	444	489	441	488	
469	11	2	-1	490	484	442	489	-1	
470	11	1	-1	443	483	490	-1	-1	
471	11	2	445	448	488	486	446	444	
472	11	2	-1	491	-1	447	450	492	449(5)
473	11	2	-1	488	448	492	452	447	
474	11	1	-1	449	-1	490	491	-1	
475	11	1	-1	450	-1	-1	491	493	
476	11	2	-1	493	-1	451	492	453	450(6)
477	11	1	-1	453	-1	-1	-1	493	
478	12	2	460	454	494	-1	-1	-1	455(1)
479	12	2	461	494	454	457	-1	-1	
480	12	2	462	457	456	495	458	-1	
481	12	2	463	-1	459	458	496	459	
482	12	2	465	-1	496	-1	459	497	
483	12	2	494	461	470	498	-1	-1	460(3)
484	12	2	495	498	469	462	499	-1	
485	12	2	496	-1	464	499	463	500	
486	12	2	466	-1	500	471	466	464	
487	12	2	497	-1	466	-1	500	465	
488	12	2	-1	473	471	501	467	468	
489	12	2	-1	502	499	468	469	501	
490	12	2	-1	469	498	474	502	-1	470(4)
491	12	2	-1	472	-1	502	475	503	474(5)
492	12	2	-1	503	-1	473	476	472	
493	12	2	-1	476	-1	-1	503	477	475(6)
494	13	2	483	479	478	504	-1	-1	
495	13	2	484	504	-1	480	505	-1	
496	13	2	485	-1	482	505	481	506	
497	13	2	487	-1	506	-1	506	482	
498	13	2	504	484	490	483	507	-1	
499	13	2	505	507	489	485	484	508	
500	13	2	506	-1	486	508	487	485	
501	13	2	-1	509	508	488	-1	489	
502	13	2	-1	489	507	491	490	509	
503	13	2	-1	492	-1	509	493	491	
504	14	2	498	495	-1	494	510	-1	
505	14	2	499	510	-1	496	495	511	
506	14	2	500	-1	497	511	497	496	
507	14	2	510	499	502	-1	498	512	
508	14	2	511	512	501	500	-1	499	
509	14	2	-1	501	512	503	-1	502	
510	15	2	507	505	-1	-1	504	513	
511	15	2	508	513	-1	506	-1	505	
512	15	2	513	508	509	-1	-1	507	
513	16	2	512	511	-1	-1	-1	510	

TABLE 9.2. $\mathcal{D}$ diagram for $E_{7,-25}$

Parameter	ℓ	d	Simples 1	2	3	4	5	6	7	Other
1	0	0	-1	-1	-1	-1	-1	-1	57	
2	0	0	-1	-1	-1	-1	-1	58	57	
3	0	0	-1	-1	-1	-1	59	58	-1	
4	0	0	-1	-1	-1	60	59	-1	-1	
5	0	0	-1	61	62	60	-1	-1	-1	
6	0	0	-1	61	63	-1	-1	-1	-1	
7	0	0	64	65	62	-1	-1	-1	-1	
8	0	0	66	65	63	67	-1	-1	-1	
9	0	0	64	68	-1	-1	-1	-1	-1	
10	0	0	66	68	-1	69	-1	-1	-1	
11	0	0	70	-1	-1	67	71	-1	-1	
12	0	0	70	-1	72	69	73	-1	-1	
13	0	0	74	-1	-1	-1	71	75	-1	
14	0	0	-1	-1	72	-1	76	-1	-1	
15	0	0	74	-1	77	-1	73	78	-1	
16	0	0	79	-1	-1	-1	-1	75	80	
17	0	0	-1	-1	77	81	76	82	-1	
18	0	0	79	-1	83	-1	-1	78	84	
19	0	0	85	-1	-1	-1	-1	-1	80	
20	0	0	-1	86	-1	81	-1	87	-1	
21	0	0	-1	-1	83	88	-1	82	89	
22	0	0	85	-1	90	-1	-1	-1	84	
23	0	0	-1	86	-1	-1	-1	91	-1	
24	0	0	-1	92	-1	88	93	87	94	
25	0	0	-1	-1	90	95	-1	-1	89	
26	0	0	-1	92	-1	-1	96	91	97	
27	0	0	-1	98	-1	-1	93	-1	99	
28	0	0	-1	100	-1	95	101	-1	94	
29	0	0	-1	98	-1	102	96	-1	103	
30	0	0	-1	100	-1	-1	104	-1	97	
31	0	0	-1	105	-1	-1	101	106	99	
32	0	0	-1	-1	107	102	-1	-1	108	
33	0	0	-1	105	-1	109	104	110	103	
34	0	0	-1	111	-1	-1	-1	106	-1	
35	0	0	112	-1	107	-1	-1	-1	113	
36	0	0	-1	-1	114	109	-1	115	108	
37	0	0	-1	111	-1	116	-1	110	-1	
38	0	0	112	-1	-1	-1	-1	-1	117	
39	0	0	118	-1	114	-1	-1	119	113	
40	0	0	-1	-1	120	116	121	115	-1	
41	0	0	118	-1	-1	-1	-1	122	117	
42	0	0	123	-1	120	-1	124	119	-1	
43	0	0	-1	-1	125	-1	121	-1	-1	
44	0	0	123	-1	-1	-1	126	122	-1	
45	0	0	127	-1	125	128	124	-1	-1	
46	0	0	127	-1	-1	129	126	-1	-1	
47	0	0	130	131	-1	128	-1	-1	-1	
48	0	0	130	132	133	129	-1	-1	-1	
49	0	0	134	131	-1	-1	-1	-1	-1	
50	0	0	134	132	135	-1	-1	-1	-1	
51	0	0	-1	136	133	-1	-1	-1	-1	
52	0	0	-1	136	135	137	-1	-1	-1	
53	0	0	-1	-1	-1	137	138	-1	-1	
54	0	0	-1	-1	-1	-1	138	139	-1	
55	0	0	-1	-1	-1	-1	-1	139	140	
56	0	0	-1	-1	-1	-1	-1	-1	140	
57	1	1	-1	-1	-1	-1	-1	141	2	1(7)
58	1	1	-1	-1	-1	-1	142	3	141	2(6)
59	1	1	-1	-1	-1	143	4	142	-1	3(5)
60	1	1	-1	144	145	5	143	-1	-1	4(4)
61	1	1	-1	6	146	144	-1	-1	-1	5(2)
62	1	1	147	146	7	145	-1	-1	-1	5(3)
63	1	1	148	146	8	149	-1	-1	-1	6(3)
64	1	1	9	150	147	-1	-1	-1	-1	7(1)
65	1	1	150	8	146	151	-1	-1	-1	7(2)
66	1	1	10	150	148	152	-1	-1	-1	8(1)
67	1	1	152	151	149	11	153	-1	-1	8(4)
68	1	1	150	10	-1	154	-1	-1	-1	9(2)
69	1	1	152	154	155	12	156	-1	-1	10(4)
70	1	1	12	-1	157	152	158	-1	-1	11(1)
71	1	1	158	-1	-1	153	13	159	-1	11(5)
72	1	1	157	-1	14	155	160	-1	-1	12(3)
73	1	1	158	-1	160	156	15	161	-1	12(5)
74	1	1	15	-1	162	-1	158	163	-1	13(1)
75	1	1	163	-1	-1	-1	159	16	164	13(6)
76	1	1	-1	-1	160	165	17	166	-1	14(5)
77	1	1	162	-1	17	167	160	168	-1	15(3)
78	1	1	163	-1	168	-1	161	18	169	15(6)
79	1	1	18	-1	170	-1	-1	163	171	16(1)
80	1	1	171	-1	-1	-1	-1	164	19	16(7)
81	1	1	-1	172	167	20	165	173	-1	17(4)
82	1	1	-1	-1	168	173	166	21	174	17(6)
83	1	1	170	-1	21	175	-1	168	176	18(3)
84	1	1	171	-1	176	-1	-1	169	22	18(7)
85	1	1	22	-1	177	-1	-1	-1	171	19(1)
86	1	1	-1	23	-1	172	-1	178	-1	20(2)
87	1	1	-1	178	-1	173	179	24	180	20(6)
88	1	1	-1	181	175	24	182	173	183	21(4)
89	1	1	-1	-1	176	183	-1	174	25	21(7)
90	1	1	177	-1	25	184	-1	-1	176	22(3)
91	1	1	-1	178	-1	-1	185	26	186	23(6)
92	1	1	-1	26	-1	181	187	178	188	24(2)
93	1	1	-1	187	-1	182	27	179	189	24(5)
94	1	1	-1	188	-1	183	189	180	28	24(7)
95	1	1	-1	190	184	28	191	-1	183	25(4)
96	1	1	-1	187	-1	192	29	185	193	26(5)
97	1	1	-1	188	-1	-1	193	186	30	26(7)
98	1	1	-1	29	-1	194	187	-1	195	27(2)
99	1	1	-1	195	-1	-1	189	196	31	27(7)
100	1	1	-1	30	-1	190	197	-1	188	28(2)
101	1	1	-1	197	-1	191	31	198	189	28(5)
102	1	1	-1	194	199	32	192	-1	200	29(4)
103	1	1	-1	195	-1	200	193	201	33	29(7)
104	1	1	-1	197	-1	202	33	203	193	30(5)
105	1	1	-1	33	-1	204	197	205	195	31(2)
106	1	1	-1	205	-1	-1	198	34	196	31(6)
107	1	1	206	-1	35	199	-1	-1	207	32(3)
108	1	1	-1	-1	207	200	-1	208	36	32(7)
109	1	1	-1	204	209	36	202	210	200	33(4)
110	1	1	-1	205	-1	210	203	37	201	33(6)
111	1	1	-1	37	-1	211	-1	205	-1	34(2)
112	1	1	38	-1	206	-1	-1	-1	212	35(1)
113	1	1	212	-1	207	-1	-1	213	39	35(7)
114	1	1	214	-1	39	209	-1	215	207	36(3)
115	1	1	-1	-1	215	210	216	40	208	36(6)
116	1	1	-1	211	217	40	218	210	-1	37(4)
117	1	1	212	-1	-1	-1	-1	219	41	38(7)
118	1	1	41	-1	214	-1	-1	220	212	39(1)
119	1	1	220	-1	215	-1	221	42	213	39(6)
120	1	1	222	-1	42	217	223	215	-1	40(3)
121	1	1	-1	-1	223	218	43	216	-1	40(5)
122	1	1	220	-1	-1	-1	224	44	219	41(6)
123	1	1	44	-1	222	-1	225	220	-1	42(1)
124	1	1	225	-1	223	226	45	221	-1	42(5)
125	1	1	227	-1	45	228	223	-1	-1	43(3)
126	1	1	225	-1	-1	229	46	224	-1	44(5)
127	1	1	46	-1	227	230	225	-1	-1	45(1)
128	1	1	230	231	228	47	226	-1	-1	45(4)
129	1	1	230	232	233	48	229	-1	-1	46(4)
130	1	1	48	234	235	230	-1	-1	-1	47(1)
131	1	1	234	49	-1	231	-1	-1	-1	47(2)
132	1	1	234	50	236	232	-1	-1	-1	48(2)
133	1	1	235	236	51	233	-1	-1	-1	48(3)
134	1	1	50	234	237	-1	-1	-1	-1	49(1)
135	1	1	237	236	52	238	-1	-1	-1	50(3)
136	1	1	-1	52	236	239	-1	-1	-1	51(2)
137	1	1	-1	239	238	53	240	-1	-1	52(4)
138	1	1	-1	-1	-1	240	54	241	-1	53(5)
139	1	1	-1	-1	-1	-1	241	55	242	54(6)
140	1	1	-1	-1	-1	-1	-1	242	56	55(7)
141	2	1	-1	-1	-1	-1	243	57	58	
142	2	1	-1	-1	-1	244	58	59	243	
143	2	1	-1	245	246	59	60	244	-1	
144	2	1	-1	60	247	61	245	-1	-1	

Parameter	ℓ	d	Simples 1	2	3	4	5	6	7	Other
145	2	1	248	247	60	62	246	-1	-1	
146	2	2	249	63	65	250	-1	-1	-1	62(2) 61(3)
147	2	1	62	249	64	248	-1	-1	-1	
148	2	1	63	249	66	251	-1	-1	-1	
149	2	1	251	252	67	63	253	-1	-1	
150	2	2	68	66	249	254	-1	-1	-1	65(1) 64(2)
151	2	1	254	67	252	65	255	-1	-1	
152	2	2	69	254	256	70	257	-1	-1	67(1) 66(4)
153	2	1	257	255	253	71	67	258	-1	
154	2	1	254	69	259	68	260	-1	-1	
155	2	1	261	259	69	72	262	-1	-1	
156	2	1	257	260	262	73	69	263	-1	
157	2	1	72	-1	70	261	264	-1	-1	
158	2	2	73	-1	264	257	74	265	-1	71(1) 70(5)
159	2	1	265	-1	-1	258	75	71	266	
160	2	2	264	-1	76	267	77	268	-1	73(3) 72(5)
161	2	1	265	-1	268	263	78	73	269	
162	2	1	77	-1	74	270	264	271	-1	
163	2	2	78	-1	271	-1	265	79	272	75(1) 74(6)
164	2	1	272	-1	-1	-1	266	80	75	
165	2	1	-1	273	274	76	81	275	-1	
166	2	1	-1	-1	268	275	82	76	276	
167	2	1	270	277	81	77	274	278	-1	
168	2	2	271	-1	82	278	268	83	279	78(3) 77(6)
169	2	1	272	-1	279	-1	269	84	78	
170	2	1	83	-1	79	280	-1	271	281	
171	2	2	84	-1	281	-1	-1	272	85	80(1) 79(7)
172	2	1	-1	81	277	86	273	282	-1	
173	2	2	-1	282	278	87	283	88	284	82(4) 81(6)
174	2	1	-1	-1	279	284	276	89	82	
175	2	1	280	285	88	83	286	278	287	
176	2	2	281	-1	89	287	-1	279	90	84(3) 83(7)
177	2	1	90	-1	85	288	-1	-1	281	
178	2	2	-1	91	-1	282	289	92	290	87(2) 86(6)
179	2	1	-1	289	-1	291	87	93	292	
180	2	1	-1	290	-1	284	292	94	87	
181	2	1	-1	88	285	92	293	282	294	
182	2	1	-1	293	286	93	88	291	295	
183	2	2	-1	294	287	94	295	284	95	89(4) 88(7)
184	2	1	288	296	95	90	297	-1	287	
185	2	1	-1	289	-1	298	91	96	299	
186	2	1	-1	290	-1	-1	299	97	91	
187	2	2	-1	96	-1	300	98	289	301	93(2) 92(5)
188	2	2	-1	97	-1	294	301	290	100	94(2) 92(7)
189	2	2	-1	301	-1	295	99	302	101	94(5) 93(7)
190	2	1	-1	95	296	100	303	-1	294	
191	2	1	-1	303	297	101	95	304	295	
192	2	1	-1	305	306	96	102	298	307	
193	2	2	-1	301	-1	307	103	308	104	97(5) 96(7)
194	2	1	-1	102	309	98	305	-1	310	
195	2	2	-1	103	-1	310	301	311	105	99(2) 98(7)
196	2	1	-1	311	-1	-1	312	99	106	
197	2	2	-1	104	-1	313	105	314	301	101(2) 100(5)
198	2	1	-1	314	-1	304	106	101	312	
199	2	1	315	309	102	107	306	-1	316	
200	2	2	-1	310	316	108	307	317	109	103(4) 102(7)
201	2	1	-1	311	-1	317	318	103	110	
202	2	1	-1	319	320	104	109	321	307	
203	2	1	-1	314	-1	321	110	104	318	
204	2	1	-1	109	322	105	319	323	310	
205	2	2	-1	110	-1	323	314	111	311	106(2) 105(6)
206	2	1	107	-1	112	315	-1	-1	324	
207	2	2	324	-1	113	316	-1	325	114	108(3) 107(7)
208	2	1	-1	-1	325	317	326	108	115	
209	2	1	327	322	109	114	320	328	316	
210	2	2	-1	323	328	115	329	116	317	110(4) 109(6)
211	2	1	-1	116	330	111	331	323	-1	
212	2	2	117	-1	324	-1	-1	332	118	113(1) 112(7)
213	2	1	332	-1	325	-1	333	113	119	
214	2	1	114	-1	118	327	-1	334	324	
215	2	2	334	-1	119	328	335	120	325	115(3) 114(6)
216	2	1	-1	-1	335	336	115	121	326	
217	2	1	337	330	116	120	338	328	-1	
218	2	1	-1	331	338	121	116	336	-1	
219	2	1	332	-1	-1	-1	339	117	122	
220	2	2	122	-1	334	-1	340	123	332	119(1) 118(6)
221	2	1	340	-1	335	341	119	124	333	
222	2	1	120	-1	123	337	342	334	-1	
223	2	2	342	-1	124	343	125	335	-1	121(3) 120(5)
224	2	1	340	-1	-1	344	122	126	339	
225	2	2	126	-1	342	345	127	340	-1	124(1) 123(5)
226	2	1	345	346	347	124	128	341	-1	
227	2	1	125	-1	127	348	342	-1	-1	
228	2	1	348	349	128	125	347	-1	-1	
229	2	1	345	350	351	126	129	344	-1	
230	2	2	129	352	353	130	345	-1	-1	128(1) 127(4)
231	2	1	352	128	349	131	346	-1	-1	
232	2	1	352	129	354	132	350	-1	-1	
233	2	1	355	354	129	133	351	-1	-1	
234	2	2	132	134	356	352	-1	-1	-1	131(1) 130(2)
235	2	1	133	356	130	355	-1	-1	-1	
236	2	2	356	135	136	357	-1	-1	-1	133(2) 132(3)
237	2	1	135	356	134	358	-1	-1	-1	
238	2	1	358	359	137	135	360	-1	-1	
239	2	1	-1	137	359	136	361	-1	-1	
240	2	1	-1	361	360	138	137	362	-1	
241	2	1	-1	-1	-1	362	139	138	363	
242	2	1	-1	-1	-1	-1	363	140	139	
243	3	1	-1	-1	-1	364	141	-1	142	
244	3	1	-1	365	366	142	-1	143	364	
245	3	1	-1	143	367	-1	144	365	-1	
246	3	1	368	367	143	-1	145	366	-1	
247	3	1	369	145	144	370	367	-1	-1	
248	3	1	145	369	-1	147	368	-1	-1	
249	3	2	146	148	150	371	-1	-1	-1	147(2)
250	3	2	371	370	370	146	372	-1	-1	
251	3	1	149	373	374	148	375	-1	-1	
252	3	1	373	149	151	370	376	-1	-1	
253	3	1	375	376	153	-1	149	377	-1	
254	3	2	154	152	378	150	379	-1	-1	151(1)
255	3	1	379	153	376	-1	151	380	-1	
256	3	2	374	378	152	374	381	-1	-1	
257	3	2	156	379	381	158	152	382	-1	153(1)
258	3	1	382	380	377	159	-1	153	383	
259	3	1	384	155	154	-1	385	-1	-1	
260	3	1	379	156	385	-1	154	386	-1	
261	3	1	155	384	374	157	387	-1	-1	
262	3	1	387	385	156	388	155	389	-1	
263	3	1	382	386	389	161	-1	156	390	
264	3	2	160	-1	158	391	162	392	-1	157(5)
265	3	2	161	-1	392	382	163	158	393	159(1)
266	3	1	393	-1	-1	383	164	-1	159	
267	3	2	391	394	388	160	388	395	-1	
268	3	2	392	-1	166	395	168	160	396	161(3)
269	3	1	393	-1	396	390	169	-1	161	
270	3	1	167	397	-1	162	398	399	-1	
271	3	2	168	-1	163	399	392	170	400	162(6)
272	3	2	169	-1	400	-1	393	171	163	164(1)
273	3	1	-1	165	401	-1	172	402	-1	
274	3	1	398	401	165	388	167	403	-1	
275	3	1	-1	402	403	166	404	165	405	
276	3	1	-1	-1	396	405	174	-1	166	
277	3	1	397	167	172	-1	401	406	-1	
278	3	2	399	406	173	168	407	175	408	167(6)
279	3	2	400	-1	174	408	396	176	168	169(3)
280	3	1	175	409	-1	170	410	399	411	
281	3	2	176	-1	171	411	-1	400	177	170(7)
282	3	2	-1	173	406	178	412	181	413	172(6)
283	3	2	-1	412	407	404	173	404	414	
284	3	2	-1	413	408	180	414	183	173	174(4)
285	3	1	409	175	181	-1	415	406	416	
286	3	1	410	415	182	-1	175	417	418	
287	3	2	411	416	183	176	418	408	184	175(7)
288	3	1	184	419	-1	177	420	-1	411	
289	3	2	-1	185	-1	421	178	187	422	179(2)
290	3	2	-1	186	-1	413	422	188	178	180(2)
291	3	1	-1	423	417	179	404	182	424	
292	3	1	-1	422	-1	424	180	425	179	
293	3	1	-1	182	415	426	181	423	427	
294	3	2	-1	183	416	188	427	413	190	181(7)
295	3	2	-1	427	418	189	183	428	191	182(7)
296	3	1	419	184	190	-1	429	-1	416	

Parameter	ℓ	d	Simples 1	2	3	4	5	6	7	Other
297	3	1	420	429	191	-1	184	430	418	
298	3	1	-1	431	432	185	-1	192	433	
299	3	1	-1	422	-1	433	186	434	185	
300	3	2	-1	426	435	187	426	421	436	
301	3	3	-1	193	-1	436	195	437	197	189(2) 188(5) 187(7)
302	3	2	-1	437	-1	428	425	189	425	
303	3	1	-1	191	429	438	190	439	427	
304	3	1	-1	439	430	198	-1	191	440	
305	3	1	-1	192	441	426	194	431	442	
306	3	1	443	441	192	-1	199	432	444	
307	3	2	-1	442	444	193	200	445	202	192(7)
308	3	2	-1	437	-1	445	434	193	434	
309	3	1	446	199	194	-1	441	-1	447	
310	3	2	-1	200	447	195	442	448	204	194(7)
311	3	2	-1	201	-1	448	449	195	205	196(2)
312	3	1	-1	449	-1	440	196	425	198	
313	3	2	-1	438	450	197	438	451	436	
314	3	2	-1	203	-1	451	205	197	449	198(2)
315	3	1	199	446	-1	206	443	-1	452	
316	3	2	452	447	200	207	444	453	209	199(7)
317	3	2	-1	448	453	208	454	200	210	201(4)
318	3	1	-1	449	-1	455	201	434	203	
319	3	1	-1	202	456	438	204	457	442	
320	3	1	458	456	202	-1	209	459	444	
321	3	1	-1	457	459	203	460	202	455	
322	3	1	461	209	204	-1	456	462	447	
323	3	2	-1	210	462	205	463	211	448	204(6)
324	3	2	207	-1	212	452	-1	464	214	206(7)
325	3	2	464	-1	213	453	465	207	215	208(3)
326	3	1	-1	-1	465	466	208	-1	216	
327	3	1	209	461	-1	214	458	467	452	
328	3	2	467	462	210	215	468	217	453	209(6)
329	3	2	-1	463	468	460	210	460	454	
330	3	1	469	217	211	-1	470	462	-1	
331	3	1	-1	218	470	-1	211	471	-1	
332	3	2	219	-1	464	-1	472	212	220	213(1)
333	3	1	472	-1	465	473	213	-1	221	
334	3	2	215	-1	220	467	474	222	464	214(6)
335	3	2	474	-1	221	475	215	223	465	216(3)
336	3	1	-1	471	476	216	460	218	466	
337	3	1	217	469	-1	222	477	467	-1	
338	3	1	477	470	218	478	217	476	-1	
339	3	1	472	-1	-1	479	219	-1	224	
340	3	2	224	-1	474	480	220	225	472	221(1)
341	3	1	480	481	482	221	-1	226	473	
342	3	2	223	-1	225	483	227	474	-1	222(5)
343	3	2	483	484	478	223	478	475	-1	
344	3	1	480	485	486	224	-1	229	479	
345	3	2	229	487	488	225	230	480	-1	226(1)
346	3	1	487	226	489	-1	231	481	-1	
347	3	1	490	489	226	478	228	482	-1	
348	3	1	228	491	492	227	490	-1	-1	
349	3	1	491	228	231	-1	489	-1	-1	
350	3	1	487	229	493	-1	232	485	-1	
351	3	1	494	493	229	-1	233	486	-1	
352	3	2	232	230	495	234	487	-1	-1	231(1)
353	3	2	492	495	230	492	488	-1	-1	
354	3	1	496	233	232	497	493	-1	-1	
355	3	1	233	496	492	235	494	-1	-1	
356	3	2	236	237	234	498	-1	-1	-1	235(2)
357	3	2	498	497	497	236	499	-1	-1	
358	3	1	238	500	-1	237	501	-1	-1	
359	3	1	500	238	239	497	502	-1	-1	
360	3	1	501	502	240	-1	238	503	-1	
361	3	1	-1	240	502	-1	239	504	-1	
362	3	1	-1	504	503	241	-1	240	505	
363	3	1	-1	-1	-1	505	242	-1	241	
364	4	1	-1	506	507	243	-1	-1	244	
365	4	1	-1	244	508	-1	-1	245	506	
366	4	1	509	508	244	-1	-1	246	507	
367	4	1	510	246	245	511	247	508	-1	
368	4	1	246	510	-1	-1	248	509	-1	
369	4	1	247	248	-1	512	510	-1	-1	
370	4	2	512	250	250	252	513	-1	-1	247(4)
371	4	2	250	512	514	249	515	-1	-1	
372	4	2	515	513	513	-1	250	516	-1	
373	4	1	252	251	517	512	518	-1	-1	
374	4	2	256	517	261	256	519	-1	-1	251(3)
375	4	1	253	518	519	-1	251	520	-1	
376	4	1	518	253	255	521	252	522	-1	
377	4	1	520	522	258	-1	-1	253	523	
378	4	2	517	256	254	514	524	-1	-1	
379	4	2	260	257	524	-1	254	525	-1	255(1)
380	4	1	525	258	522	-1	-1	255	526	
381	4	2	519	524	257	527	256	528	-1	
382	4	2	263	525	528	265	-1	257	529	258(1)
383	4	1	529	526	523	266	-1	-1	258	
384	4	1	259	261	517	-1	530	-1	-1	
385	4	1	530	262	260	531	259	532	-1	
386	4	1	525	263	532	-1	-1	260	533	
387	4	1	262	530	519	534	261	535	-1	
388	4	2	534	536	267	274	267	537	-1	262(4)
389	4	1	535	532	263	537	-1	262	538	
390	4	1	529	533	538	269	-1	-1	263	
391	4	2	267	539	527	264	534	540	-1	
392	4	2	268	-1	265	540	271	264	541	
393	4	2	269	-1	541	529	272	-1	265	266(1)
394	4	2	539	267	536	-1	536	542	-1	
395	4	2	540	542	537	268	543	267	544	
396	4	2	541	-1	276	544	279	-1	268	269(3)
397	4	1	277	270	-1	-1	545	546	-1	
398	4	1	274	545	-1	534	270	547	-1	
399	4	2	278	546	-1	271	548	280	549	270(6)
400	4	2	279	-1	272	549	541	281	271	
401	4	1	545	274	273	550	277	551	-1	
402	4	1	-1	275	551	-1	552	273	553	
403	4	1	547	551	275	537	554	274	555	
404	4	2	-1	552	554	283	291	283	556	275(5)
405	4	1	-1	553	555	276	556	-1	275	
406	4	2	546	278	282	-1	557	285	558	277(6)
407	4	2	548	557	283	543	278	554	559	
408	4	2	549	558	284	279	559	287	278	
409	4	1	285	280	-1	-1	560	546	561	
410	4	1	286	560	-1	-1	280	562	563	
411	4	2	287	561	-1	281	563	549	288	280(7)
412	4	2	-1	283	557	564	282	552	565	
413	4	2	-1	284	558	290	565	294	282	
414	4	2	-1	565	559	556	284	566	283	
415	4	1	560	286	293	567	285	568	569	
416	4	2	561	287	294	-1	569	558	296	285(7)
417	4	1	562	568	291	-1	554	286	570	
418	4	2	563	569	295	-1	287	571	297	286(7)
419	4	1	296	288	-1	-1	572	-1	561	
420	4	1	297	572	-1	-1	288	573	563	
421	4	2	-1	574	575	289	564	300	576	
422	4	2	-1	299	-1	576	290	577	289	292(2)
423	4	1	-1	291	568	574	552	293	578	
424	4	1	-1	578	570	292	556	579	291	
425	4	2	-1	577	-1	579	302	312	302	292(6)
426	4	2	-1	300	580	305	300	574	581	293(4)
427	4	2	-1	295	569	581	294	582	303	293(7)
428	4	2	-1	582	571	302	566	295	579	
429	4	1	572	297	303	583	296	584	569	
430	4	1	573	584	304	-1	-1	297	585	
431	4	1	-1	298	586	574	-1	305	587	
432	4	1	588	586	298	-1	-1	306	589	
433	4	1	-1	587	589	299	-1	590	298	
434	4	2	-1	577	-1	590	308	318	308	299(6)
435	4	2	591	580	300	-1	580	575	592	
436	4	3	-1	581	592	301	581	593	313	300(7)
437	4	3	-1	308	-1	593	577	301	577	302(2)
438	4	2	-1	313	594	319	313	595	581	303(4)
439	4	1	-1	304	584	595	-1	303	596	
440	4	1	-1	596	585	312	-1	579	304	
441	4	1	597	306	305	598	309	586	599	
442	4	2	-1	307	599	581	310	600	319	305(7)
443	4	1	306	597	-1	-1	315	588	601	
444	4	2	601	599	307	-1	316	602	320	306(7)
445	4	2	-1	600	602	308	603	307	590	
446	4	1	309	315	-1	-1	597	-1	604	
447	4	2	604	316	310	-1	599	605	322	309(7)
448	4	2	-1	317	605	311	606	310	323	

Parameter	ℓ	d	Simples 1	2	3	4	5	6	7	Other
449	4	2	-1	318	-1	607	311	577	314	312(2)
450	4	2	608	594	313	-1	594	609	592	
451	4	2	-1	595	609	314	610	313	607	
452	4	2	316	604	-1	324	601	611	327	315(7)
453	4	2	611	605	317	325	612	316	328	
454	4	2	-1	606	612	613	317	603	329	
455	4	1	-1	614	615	318	613	590	321	
456	4	1	616	320	319	617	322	618	599	
457	4	1	-1	321	618	595	619	319	614	
458	4	1	320	616	-1	-1	327	620	601	
459	4	1	620	618	321	-1	621	320	615	
460	4	2	-1	619	621	329	336	329	613	321(5)
461	4	1	322	327	-1	-1	616	622	604	
462	4	2	622	328	323	-1	623	330	605	322(6)
463	4	2	-1	329	623	610	323	619	606	
464	4	2	325	-1	332	611	624	324	334	
465	4	2	624	-1	333	625	325	-1	335	326(3)
466	4	1	-1	626	627	326	613	-1	336	
467	4	2	328	622	-1	334	628	337	611	327(6)
468	4	2	628	623	329	629	328	621	612	
469	4	1	330	337	-1	-1	630	622	-1	
470	4	1	630	338	331	631	330	632	-1	
471	4	1	-1	336	632	-1	619	331	626	
472	4	2	339	-1	624	633	332	-1	340	333(1)
473	4	1	633	634	635	333	-1	-1	341	
474	4	2	335	-1	340	636	334	342	624	
475	4	2	636	637	638	335	629	343	625	
476	4	1	639	632	336	638	621	338	627	
477	4	1	338	630	-1	640	337	639	-1	
478	4	2	640	641	343	347	343	638	-1	338(4)
479	4	1	633	642	643	339	-1	-1	344	
480	4	2	344	644	645	340	-1	345	633	341(1)
481	4	1	644	341	646	-1	-1	346	634	
482	4	1	647	646	341	638	-1	347	635	
483	4	2	343	648	649	342	640	636	-1	
484	4	2	648	343	641	-1	641	637	-1	
485	4	1	644	344	650	-1	-1	350	642	
486	4	1	651	650	344	-1	-1	351	643	
487	4	2	350	345	652	-1	352	644	-1	346(1)
488	4	2	653	652	345	649	353	645	-1	
489	4	1	654	347	346	655	349	646	-1	
490	4	1	347	654	653	640	348	647	-1	
491	4	1	349	348	656	-1	654	-1	-1	
492	4	2	353	656	355	353	653	-1	-1	348(3)
493	4	1	657	351	350	658	354	650	-1	
494	4	1	351	657	653	-1	355	651	-1	
495	4	2	656	353	352	659	652	-1	-1	
496	4	1	354	355	656	660	657	-1	-1	
497	4	2	660	357	357	359	661	-1	-1	354(4)
498	4	2	357	660	659	356	662	-1	-1	
499	4	2	662	661	661	-1	357	663	-1	
500	4	1	359	358	-1	660	664	-1	-1	
501	4	1	360	664	-1	-1	358	665	-1	
502	4	1	664	360	361	666	359	667	-1	
503	4	1	665	667	362	-1	-1	360	668	
504	4	1	-1	362	667	-1	-1	361	669	
505	4	1	-1	669	668	363	-1	-1	362	
506	5	1	-1	364	670	-1	-1	-1	365	
507	5	1	671	670	364	-1	-1	-1	366	
508	5	1	672	366	365	673	-1	367	670	
509	5	1	366	672	-1	-1	-1	368	671	
510	5	1	367	368	-1	674	369	672	-1	
511	5	1	674	-1	-1	367	675	673	-1	
512	5	2	370	371	676	373	677	-1	-1	369(4)
513	5	2	677	372	372	675	370	678	-1	
514	5	2	676	676	371	378	679	-1	-1	
515	5	2	372	677	679	-1	371	680	-1	
516	5	2	680	678	678	-1	-1	372	681	
517	5	2	378	374	384	676	682	-1	-1	373(3)
518	5	1	376	375	682	683	373	684	-1	
519	5	2	381	682	387	685	374	686	-1	375(3)
520	5	1	377	684	686	-1	-1	375	687	
521	5	1	683	-1	-1	376	675	688	-1	
522	5	1	684	377	380	688	-1	376	689	
523	5	1	687	689	383	-1	-1	-1	377	
524	5	2	682	381	379	690	378	691	-1	
525	5	2	386	382	691	-1	-1	379	692	380(1)
526	5	1	692	383	689	-1	-1	-1	380	
527	5	2	685	693	391	381	685	694	-1	
528	5	2	686	691	382	694	-1	381	695	
529	5	2	390	692	695	393	-1	-1	382	383(1)
530	5	1	385	387	682	696	384	697	-1	
531	5	1	696	698	-1	385	-1	699	-1	
532	5	1	697	389	386	699	-1	385	700	
533	5	1	692	390	700	-1	-1	-1	386	
534	5	2	388	701	685	398	391	702	-1	387(4)
535	5	1	389	697	686	702	-1	387	703	
536	5	2	701	388	394	698	394	704	-1	
537	5	2	702	704	395	403	705	388	706	389(4)
538	5	1	703	700	390	706	-1	-1	389	
539	5	2	394	391	693	-1	701	707	-1	
540	5	2	395	707	694	392	708	391	709	
541	5	2	396	-1	393	709	400	-1	392	
542	5	2	707	395	704	-1	710	394	711	
543	5	2	708	710	705	407	395	705	712	
544	5	2	709	711	706	396	712	-1	395	
545	5	1	401	398	-1	713	397	714	-1	
546	5	2	406	399	-1	-1	715	409	716	397(6)
547	5	1	403	714	-1	702	717	398	718	
548	5	2	407	715	-1	708	399	717	719	
549	5	2	408	716	-1	400	719	411	399	
550	5	1	713	698	-1	401	-1	720	-1	
551	5	1	714	403	402	720	721	401	722	
552	5	2	-1	404	721	723	423	412	724	402(5)
553	5	1	-1	405	722	-1	724	-1	402	
554	5	2	717	721	404	705	417	407	725	403(5)
555	5	1	718	722	405	706	725	-1	403	
556	5	2	-1	724	725	414	424	726	404	405(5)
557	5	2	715	407	412	727	406	721	728	
558	5	2	716	408	413	-1	728	416	406	
559	5	2	719	728	414	712	408	729	407	
560	5	1	415	410	-1	730	409	731	732	
561	5	2	416	411	-1	-1	732	716	419	409(7)
562	5	1	417	731	-1	-1	717	410	733	
563	5	2	418	732	-1	-1	411	734	420	410(7)
564	5	2	-1	723	735	412	421	723	736	
565	5	2	-1	414	728	736	413	737	412	
566	5	2	-1	737	729	726	428	414	726	
567	5	1	730	-1	738	415	-1	739	740	
568	5	1	731	417	423	739	721	415	741	
569	5	2	732	418	427	740	416	742	429	415(7)
570	5	1	733	741	424	-1	725	743	417	
571	5	2	734	742	428	-1	729	418	743	
572	5	1	429	420	-1	744	419	745	732	
573	5	1	430	745	-1	-1	-1	420	746	
574	5	2	-1	421	747	431	723	426	748	423(4)
575	5	2	749	747	421	-1	735	435	750	
576	5	2	-1	748	750	422	736	751	421	
577	5	3	-1	434	-1	751	437	449	437	425(2) 422(6)
578	5	1	-1	424	741	748	724	752	423	
579	5	2	-1	752	743	425	726	440	428	424(6)
580	5	2	753	435	426	738	435	747	754	
581	5	3	-1	436	754	442	436	755	438	427(4) 426(7)
582	5	2	-1	428	742	755	737	427	752	
583	5	1	744	-1	756	429	-1	757	740	
584	5	1	745	430	439	757	-1	429	758	
585	5	1	746	758	440	-1	-1	743	430	
586	5	1	759	432	431	760	-1	441	761	
587	5	1	-1	433	761	748	-1	762	431	
588	5	1	432	759	-1	-1	-1	443	763	
589	5	1	763	761	433	-1	-1	764	432	
590	5	2	-1	762	764	434	765	455	445	433(6)
591	5	2	435	753	-1	-1	753	749	766	
592	5	3	766	754	436	-1	754	767	450	435(7)
593	5	3	-1	755	767	437	768	436	751	
594	5	2	769	450	438	756	450	770	754	
595	5	2	-1	451	770	457	771	438	772	439(4)
596	5	1	-1	440	758	772	-1	752	439	
597	5	1	441	443	-1	773	446	759	774	
598	5	1	773	-1	738	441	-1	760	775	
599	5	2	774	444	442	775	447	776	456	441(7)
600	5	2	-1	445	776	755	777	442	762	

Parameter	ℓ	d	Simples 1	2	3	4	5	6	7	Other
601	5	2	444	774	-1	-1	452	778	458	443(7)
602	5	2	778	776	445	-1	779	444	764	
603	5	2	-1	777	779	765	445	454	765	
604	5	2	447	452	-1	-1	774	780	461	446(7)
605	5	2	780	453	448	-1	781	447	462	
606	5	2	-1	454	781	782	448	777	463	
607	5	2	-1	772	783	449	782	751	451	
608	5	2	450	769	-1	-1	769	784	766	
609	5	2	784	770	451	-1	785	450	783	
610	5	2	-1	771	785	463	451	771	782	
611	5	2	453	780	-1	464	786	452	467	
612	5	2	786	781	454	787	453	779	468	
613	5	2	-1	788	789	454	466	765	460	455(5)
614	5	1	-1	455	790	772	788	762	457	
615	5	1	791	790	455	-1	789	764	459	
616	5	1	456	458	-1	792	461	793	774	
617	5	1	792	-1	756	456	-1	794	775	
618	5	1	793	459	457	794	795	456	790	
619	5	2	-1	460	795	771	471	463	788	457(5)
620	5	1	459	793	-1	-1	796	458	791	
621	5	2	796	795	460	797	476	468	789	459(5)
622	5	2	462	467	-1	-1	798	469	780	461(6)
623	5	2	798	468	463	799	462	795	781	
624	5	2	465	-1	472	800	464	-1	474	
625	5	2	800	801	802	465	787	-1	475	
626	5	1	-1	466	803	-1	788	-1	471	
627	5	1	804	803	466	802	789	-1	476	
628	5	2	468	798	-1	805	467	796	786	
629	5	2	805	806	797	468	475	797	787	
630	5	1	470	477	-1	807	469	808	-1	
631	5	1	807	809	-1	470	-1	810	-1	
632	5	1	808	476	471	810	795	470	803	
633	5	2	479	811	812	472	-1	-1	480	473(1)
634	5	1	811	473	813	-1	-1	-1	481	
635	5	1	814	813	473	802	-1	-1	482	
636	5	2	475	815	816	474	805	483	800	
637	5	2	815	475	817	-1	806	484	801	
638	5	2	818	817	475	482	797	478	802	476(4)
639	5	1	476	808	-1	818	796	477	804	
640	5	2	478	819	820	490	483	818	-1	477(4)
641	5	2	819	478	484	809	484	817	-1	
642	5	1	811	479	821	-1	-1	-1	485	
643	5	1	822	821	479	-1	-1	-1	486	
644	5	2	485	480	823	-1	-1	487	811	481(1)
645	5	2	824	823	480	816	-1	488	812	
646	5	1	825	482	481	826	-1	489	813	
647	5	1	482	825	824	818	-1	490	814	
648	5	2	484	483	827	-1	819	815	-1	
649	5	2	820	827	483	488	820	816	-1	
650	5	1	828	486	485	829	-1	493	821	
651	5	1	486	828	824	-1	-1	494	822	
652	5	2	830	488	487	831	495	823	-1	
653	5	2	488	830	494	820	492	824	-1	490(3)
654	5	1	489	490	830	832	491	825	-1	
655	5	1	832	809	-1	489	-1	826	-1	
656	5	2	495	492	496	833	830	-1	-1	491(3)
657	5	1	493	494	830	834	496	828	-1	
658	5	1	834	-1	-1	493	835	829	-1	
659	5	2	833	833	498	495	836	-1	-1	
660	5	2	497	498	833	500	837	-1	-1	496(4)
661	5	2	837	499	499	835	497	838	-1	
662	5	2	499	837	836	-1	498	839	-1	
663	5	2	839	838	838	-1	-1	499	840	
664	5	1	502	501	-1	841	500	842	-1	
665	5	1	503	842	-1	-1	-1	501	843	
666	5	1	841	-1	-1	502	835	844	-1	
667	5	1	842	503	504	844	-1	502	845	
668	5	1	843	845	505	-1	-1	-1	503	
669	5	1	-1	505	845	-1	-1	-1	504	
670	6	1	846	507	506	847	-1	-1	508	
671	6	1	507	846	-1	-1	-1	-1	509	
672	6	1	508	509	-1	848	-1	510	846	
673	6	1	848	-1	-1	508	849	511	847	
674	6	1	511	-1	850	510	851	848	-1	
675	6	2	851	-1	-1	513	521	852	-1	511(5)
676	6	2	514	514	512	517	853	-1	-1	
677	6	2	513	515	853	851	512	854	-1	
678	6	2	854	516	516	852	-1	513	855	
679	6	2	853	853	515	856	514	857	-1	
680	6	2	516	854	857	-1	-1	515	858	
681	6	2	858	855	855	-1	-1	-1	516	
682	6	2	524	519	530	859	517	860	-1	518(3)
683	6	1	521	-1	861	518	851	862	-1	
684	6	1	522	520	860	862	-1	518	863	
685	6	2	527	864	534	519	527	865	-1	
686	6	2	528	860	535	865	-1	519	866	520(3)
687	6	1	523	863	866	-1	-1	-1	520	
688	6	1	862	-1	-1	522	867	521	868	
689	6	1	863	523	526	868	-1	-1	522	
690	6	2	859	869	-1	524	856	870	-1	
691	6	2	860	528	525	870	-1	524	871	
692	6	2	533	529	871	-1	-1	-1	525	526(1)
693	6	2	864	527	539	869	864	872	-1	
694	6	2	865	872	540	528	873	527	874	
695	6	2	866	871	529	874	-1	-1	528	
696	6	1	531	875	876	530	-1	877	-1	
697	6	1	532	535	860	877	-1	530	878	
698	6	2	875	550	-1	536	-1	879	-1	531(2)
699	6	1	877	879	-1	532	880	531	881	
700	6	1	878	538	533	881	-1	-1	532	
701	6	2	536	534	864	875	539	882	-1	
702	6	2	537	882	865	547	883	534	884	535(4)
703	6	1	538	878	866	884	-1	-1	535	
704	6	2	882	537	542	879	885	536	886	
705	6	2	883	885	543	554	537	543	887	
706	6	2	884	886	544	555	887	-1	537	538(4)
707	6	2	542	540	872	-1	888	539	889	
708	6	2	543	888	873	548	540	883	890	
709	6	2	544	889	874	541	890	-1	540	
710	6	2	888	543	885	891	542	885	892	
711	6	2	889	544	886	-1	892	-1	542	
712	6	2	890	892	887	559	544	893	543	
713	6	1	550	875	894	545	-1	895	-1	
714	6	1	551	547	-1	895	896	545	897	
715	6	2	557	548	-1	898	546	896	899	
716	6	2	558	549	-1	-1	899	561	546	
717	6	2	554	896	-1	883	562	548	900	547(5)
718	6	1	555	897	-1	884	900	-1	547	
719	6	2	559	899	-1	890	549	901	548	
720	6	1	895	879	-1	551	902	550	903	
721	6	2	896	554	552	904	568	557	905	551(5)
722	6	1	897	555	553	903	905	-1	551	
723	6	2	-1	564	906	552	574	564	907	
724	6	2	-1	556	905	907	578	908	552	553(5)
725	6	2	900	905	556	887	570	909	554	555(5)
726	6	2	-1	908	909	566	579	556	566	
727	6	2	898	891	910	557	-1	904	911	
728	6	2	899	559	565	911	558	912	557	
729	6	2	901	912	566	893	571	559	909	
730	6	1	567	-1	913	560	-1	914	915	
731	6	1	568	562	-1	914	896	560	916	
732	6	2	569	563	-1	915	561	917	572	560(7)
733	6	1	570	916	-1	-1	900	918	562	
734	6	2	571	917	-1	-1	901	563	918	
735	6	2	919	906	564	910	575	906	920	
736	6	2	-1	907	920	565	576	921	564	
737	6	2	-1	566	912	921	582	565	908	
738	6	2	922	-1	598	580	-1	923	924	567(3)
739	6	1	914	-1	923	568	925	567	926	
740	6	2	915	-1	924	569	-1	927	583	567(7)
741	6	1	916	570	578	926	905	928	568	
742	6	2	917	571	582	927	912	569	928	
743	6	2	918	928	579	-1	909	585	571	570(6)
744	6	1	583	-1	929	572	-1	930	915	
745	6	1	584	573	-1	930	-1	572	931	
746	6	1	585	931	-1	-1	-1	918	573	
747	6	2	932	575	574	923	906	580	933	
748	6	2	-1	576	933	587	907	934	574	578(4)
749	6	2	575	932	-1	-1	919	591	935	
750	6	2	935	933	576	-1	920	936	575	
751	6	3	-1	934	936	577	937	607	593	576(6)
752	6	2	-1	579	928	934	908	596	582	578(6)

Parameter	ℓ	d	Simples 1	2	3	4	5	6	7	Other
753	6	2	580	591	-1	922	591	932	938	
754	6	3	938	592	581	924	592	939	594	580(7)
755	6	3	-1	593	939	600	940	581	934	582(4)
756	6	2	941	-1	617	594	-1	942	924	583(3)
757	6	1	930	-1	942	584	943	583	944	
758	6	1	931	585	596	944	-1	928	584	
759	6	1	586	588	-1	945	-1	597	946	
760	6	1	945	-1	923	586	947	598	948	
761	6	1	946	589	587	948	-1	949	586	
762	6	2	-1	590	949	934	950	614	600	587(6)
763	6	1	589	946	-1	-1	-1	951	588	
764	6	2	951	949	590	-1	952	615	602	589(6)
765	6	2	-1	950	952	603	590	613	603	
766	6	3	592	938	-1	-1	938	953	608	591(7)
767	6	3	953	939	593	-1	954	592	936	
768	6	3	-1	940	954	937	593	940	937	
769	6	2	594	608	-1	941	608	955	938	
770	6	2	955	609	595	942	956	594	957	
771	6	2	-1	610	956	619	595	610	958	
772	6	2	-1	607	957	614	958	934	595	596(4)
773	6	1	598	-1	959	597	-1	945	960	
774	6	2	599	601	-1	960	604	961	616	597(7)
775	6	2	960	-1	924	599	-1	962	617	598(7)
776	6	2	961	602	600	962	963	599	949	
777	6	2	-1	603	963	964	600	606	950	
778	6	2	602	961	-1	-1	965	601	951	
779	6	2	965	963	603	966	602	612	952	
780	6	2	605	611	-1	-1	967	604	622	
781	6	2	967	612	606	968	605	963	623	
782	6	2	-1	958	969	606	607	964	610	
783	6	2	970	957	607	-1	969	936	609	
784	6	2	609	955	-1	-1	971	608	970	
785	6	2	971	956	610	972	609	956	969	
786	6	2	612	967	-1	973	611	965	628	
787	6	2	973	974	975	612	625	966	629	
788	6	2	-1	613	976	958	626	950	619	614(5)
789	6	2	977	976	613	975	627	952	621	615(5)
790	6	1	978	615	614	979	976	949	618	
791	6	1	615	978	-1	-1	977	951	620	
792	6	1	617	-1	980	616	-1	981	960	
793	6	1	618	620	-1	981	982	616	978	
794	6	1	981	-1	942	618	983	617	979	
795	6	2	982	621	619	984	632	623	976	618(5)
796	6	2	621	982	-1	985	639	628	977	620(5)
797	6	2	985	986	629	621	638	629	975	
798	6	2	623	628	-1	987	622	982	967	
799	6	2	987	988	972	623	-1	984	968	
800	6	2	625	989	990	624	973	-1	636	
801	6	2	989	625	991	-1	974	-1	637	
802	6	2	992	991	625	635	975	-1	638	627(4)
803	6	1	993	627	626	994	976	-1	632	
804	6	1	627	993	-1	992	977	-1	639	
805	6	2	629	995	996	628	636	985	973	
806	6	2	995	629	986	988	637	986	974	
807	6	1	631	997	998	630	-1	999	-1	
808	6	1	632	639	-1	999	982	630	993	
809	6	2	997	655	-1	641	-1	1000	-1	631(2)
810	6	1	999	1000	-1	632	1001	631	994	
811	6	2	642	633	1002	-1	-1	-1	644	634(1)
812	6	2	1003	1002	633	990	-1	-1	645	
813	6	1	1004	635	634	1005	-1	-1	646	
814	6	1	635	1004	1003	992	-1	-1	647	
815	6	2	637	636	1006	-1	995	648	989	
816	6	2	1007	1006	636	645	996	649	990	
817	6	2	1008	638	637	1000	986	641	991	
818	6	2	638	1008	1007	647	985	640	992	639(4)
819	6	2	641	640	1009	997	648	1008	-1	
820	6	2	649	1009	640	653	649	1007	-1	
821	6	1	1010	643	642	1011	-1	-1	650	
822	6	1	643	1010	1003	-1	-1	-1	651	
823	6	2	1012	645	644	1013	-1	652	1002	
824	6	2	645	1012	651	1007	-1	653	1003	647(3)
825	6	1	646	647	1012	1014	-1	654	1004	
826	6	1	1014	1000	-1	646	1015	655	1005	
827	6	2	1009	649	648	1016	1009	1006	-1	
828	6	1	650	651	1012	1017	-1	657	1010	
829	6	1	1017	-1	-1	650	1018	658	1011	
830	6	2	652	653	657	1019	656	1012	-1	654(3)
831	6	2	1019	1016	-1	652	1020	1013	-1	
832	6	1	655	997	1021	654	-1	1014	-1	
833	6	2	659	659	660	656	1022	-1	-1	
834	6	1	658	-1	1023	657	1024	1017	-1	
835	6	2	1024	-1	-1	661	666	1025	-1	658(5)
836	6	2	1022	1022	662	1020	659	1026	-1	
837	6	2	661	662	1022	1024	660	1027	-1	
838	6	2	1027	663	663	1025	-1	661	1028	
839	6	2	663	1027	1026	-1	-1	662	1029	
840	6	2	1029	1028	1028	-1	-1	-1	663	
841	6	1	666	-1	1030	664	1024	1031	-1	
842	6	1	667	665	-1	1031	-1	664	1032	
843	6	1	668	1032	-1	-1	-1	-1	665	
844	6	1	1031	-1	-1	667	1033	666	1034	
845	6	1	1032	668	669	1034	-1	-1	667	
846	7	1	670	671	-1	1035	-1	-1	672	
847	7	1	1035	-1	-1	670	1036	-1	673	
848	7	1	673	-1	1037	672	1038	674	1035	
849	7	1	1038	-1	-1	-1	673	1039	1036	
850	7	1	-1	-1	674	-1	1040	1037	-1	
851	7	2	675	-1	1040	677	683	1041	-1	674(5)
852	7	2	1041	-1	-1	678	1039	675	1042	
853	7	2	679	679	677	1043	676	1044	-1	
854	7	2	678	680	1044	1041	-1	677	1045	
855	7	2	1045	681	681	1042	-1	-1	678	
856	7	2	1043	1046	-1	679	690	1047	-1	
857	7	2	1044	1044	680	1047	-1	679	1048	
858	7	2	681	1045	1048	-1	-1	-1	680	
859	7	2	690	1049	1050	682	1043	1051	-1	
860	7	2	691	686	697	1051	-1	682	1052	684(3)
861	7	1	-1	-1	683	1050	1040	1053	-1	
862	7	1	688	-1	1053	684	1054	683	1055	
863	7	1	689	687	1052	1055	-1	-1	684	
864	7	2	693	685	701	1049	693	1056	-1	
865	7	2	694	1056	702	686	1057	685	1058	
866	7	2	695	1052	703	1058	-1	-1	686	687(3)
867	7	1	1054	-1	-1	-1	688	1039	1059	
868	7	1	1055	-1	-1	689	1059	-1	688	
869	7	2	1049	690	-1	693	1046	1060	-1	
870	7	2	1051	1060	-1	691	1061	690	1062	
871	7	2	1052	695	692	1062	-1	-1	691	
872	7	2	1056	694	707	1060	1063	693	1064	
873	7	2	1057	1063	708	-1	694	1057	1065	
874	7	2	1058	1064	709	695	1065	-1	694	
875	7	2	698	713	1066	701	-1	1067	-1	696(2)
876	7	1	-1	1066	696	1050	-1	1068	-1	
877	7	1	699	1067	1068	697	1069	696	1070	
878	7	1	700	703	1052	1070	-1	-1	697	
879	7	2	1067	720	-1	704	1071	698	1072	699(2)
880	7	1	1069	1071	-1	-1	699	-1	1073	
881	7	1	1070	1072	-1	700	1073	-1	699	
882	7	2	704	702	1056	1067	1074	701	1075	
883	7	2	705	1074	1057	717	702	708	1076	
884	7	2	706	1075	1058	718	1076	-1	702	703(4)
885	7	2	1074	705	710	1077	704	710	1078	
886	7	2	1075	706	711	1072	1078	-1	704	
887	7	2	1076	1078	712	725	706	1079	705	
888	7	2	710	708	1063	1080	707	1074	1081	
889	7	2	711	709	1064	-1	1081	-1	707	
890	7	2	712	1081	1065	719	709	1082	708	
891	7	2	1080	727	1083	710	-1	1077	1084	
892	7	2	1081	712	1078	1084	711	1085	710	
893	7	2	1082	1085	1079	729	-1	712	1079	
894	7	1	-1	1066	713	-1	-1	1086	-1	
895	7	1	720	1067	1086	714	1087	713	1088	
896	7	2	721	717	-1	1089	731	715	1090	714(5)
897	7	1	722	718	-1	1088	1090	-1	714	
898	7	2	727	1080	1091	715	-1	1089	1092	
899	7	2	728	719	-1	1092	716	1093	715	
900	7	2	725	1090	-1	1076	733	1094	717	718(5)
901	7	2	729	1093	-1	1082	734	719	1094	
902	7	1	1087	1071	-1	1095	720	-1	1096	
903	7	1	1088	1072	-1	722	1096	-1	720	
904	7	2	1089	1077	1097	721	1095	727	1098	

Parameter	ℓ	d	Simples 1	2	3	4	5	6	7	Other
905	7	2	1090	725	724	1098	741	1099	721	722(5)
906	7	2	1100	735	723	1097	747	735	1101	
907	7	2	-1	736	1101	724	748	1102	723	
908	7	2	-1	726	1099	1102	752	724	737	
909	7	2	1094	1099	726	1079	743	725	729	
910	7	2	1103	1083	727	735	-1	1097	1104	
911	7	2	1092	1084	1104	728	-1	1105	727	
912	7	2	1093	729	737	1105	742	728	1099	
913	7	1	1106	-1	730	-1	-1	1107	1108	
914	7	1	739	-1	1107	731	1109	730	1110	
915	7	2	740	-1	1108	732	-1	1111	744	730(7)
916	7	1	741	733	-1	1110	1090	1112	731	
917	7	2	742	734	-1	1111	1093	732	1112	
918	7	2	743	1112	-1	-1	1094	746	734	733(6)
919	7	2	735	1100	-1	1103	749	1100	1113	
920	7	2	1113	1101	736	1104	750	1114	735	
921	7	2	-1	1102	1114	737	1115	736	1102	
922	7	2	738	-1	1106	753	-1	1116	1117	
923	7	2	1116	-1	760	747	1118	738	1119	739(3)
924	7	3	1117	-1	775	754	-1	1120	756	740(3) 738(7)
925	7	1	1109	-1	1118	1095	739	-1	1121	
926	7	1	1110	-1	1119	741	1121	1122	739	
927	7	2	1111	-1	1120	742	1123	740	1122	
928	7	2	1112	743	752	1122	1099	758	742	741(6)
929	7	1	1124	-1	744	-1	-1	1125	1108	
930	7	1	757	-1	1125	745	1126	744	1127	
931	7	1	758	746	-1	1127	-1	1112	745	
932	7	2	747	749	-1	1116	1100	753	1128	
933	7	2	1128	750	748	1119	1101	1129	747	
934	7	3	-1	751	1129	762	1130	772	755	752(4) 748(6)
935	7	2	750	1128	-1	-1	1113	1131	749	
936	7	3	1131	1129	751	-1	1132	783	767	750(6)
937	7	3	-1	1130	1132	768	751	1115	768	
938	7	3	754	766	-1	1117	766	1133	769	753(7)
939	7	3	1133	767	755	1120	1134	754	1129	
940	7	3	-1	768	1134	1115	755	768	1130	
941	7	2	756	-1	1124	769	-1	1135	1117	
942	7	2	1135	-1	794	770	1136	756	1137	757(3)
943	7	1	1126	-1	1136	-1	757	-1	1138	
944	7	1	1127	-1	1137	758	1138	1122	757	
945	7	1	760	-1	1139	759	1140	773	1141	
946	7	1	761	763	-1	1141	-1	1142	759	
947	7	1	1140	-1	1118	-1	760	-1	1143	
948	7	1	1141	-1	1119	761	1143	1144	760	
949	7	2	1142	764	762	1144	1145	790	776	761(6)
950	7	2	-1	765	1145	1146	762	788	777	
951	7	2	764	1142	-1	-1	1147	791	778	763(6)
952	7	2	1147	1145	765	1148	764	789	779	
953	7	3	767	1133	-1	-1	1149	766	1131	
954	7	3	1149	1134	768	1150	767	1134	1132	
955	7	2	770	784	-1	1135	1151	769	1152	
956	7	2	1151	785	771	1153	770	785	1154	
957	7	2	1152	783	772	1137	1154	1129	770	
958	7	2	-1	782	1154	788	772	1146	771	
959	7	1	1106	-1	773	-1	-1	1139	1155	
960	7	2	775	-1	1155	774	-1	1156	792	773(7)
961	7	2	776	778	-1	1156	1157	774	1142	
962	7	2	1156	-1	1120	776	1158	775	1144	
963	7	2	1157	779	777	1159	776	781	1145	
964	7	2	-1	1146	1160	777	1115	782	1146	
965	7	2	779	1157	-1	1161	778	786	1147	
966	7	2	1161	1162	1148	779	-1	787	1148	
967	7	2	781	786	-1	1163	780	1157	798	
968	7	2	1163	1164	1165	781	-1	1159	799	
969	7	2	1166	1154	782	1165	783	1160	785	
970	7	2	783	1152	-1	-1	1166	1131	784	
971	7	2	785	1151	-1	1167	784	1151	1166	
972	7	2	1167	1168	799	785	-1	1153	1165	
973	7	2	787	1169	1170	786	800	1161	805	
974	7	2	1169	787	1171	1164	801	1162	806	
975	7	2	1172	1171	787	789	802	1148	797	
976	7	2	1173	789	788	1174	803	1145	795	790(5)
977	7	2	789	1173	-1	1172	804	1147	796	791(5)
978	7	1	790	791	-1	1175	1173	1142	793	
979	7	1	1175	-1	1137	790	1176	1144	794	
980	7	1	1124	-1	792	-1	-1	1177	1155	
981	7	1	794	-1	1177	793	1178	792	1175	
982	7	2	795	796	-1	1179	808	798	1173	793(5)
983	7	1	1178	-1	1136	1180	794	-1	1176	
984	7	2	1179	1181	1153	795	1180	799	1174	
985	7	2	797	1182	1183	796	818	805	1172	
986	7	2	1182	797	806	1181	817	806	1171	
987	7	2	799	1184	1185	798	-1	1179	1163	
988	7	2	1184	799	1168	806	-1	1181	1164	
989	7	2	801	800	1186	-1	1169	-1	815	
990	7	2	1187	1186	800	812	1170	-1	816	
991	7	2	1188	802	801	1189	1171	-1	817	
992	7	2	802	1188	1187	814	1172	-1	818	804(4)
993	7	1	803	804	-1	1190	1173	-1	808	
994	7	1	1190	1189	-1	803	1191	-1	810	
995	7	2	806	805	1192	1184	815	1182	1169	
996	7	2	1183	1192	805	-1	816	1183	1170	
997	7	2	809	832	1193	819	-1	1194	-1	807(2)
998	7	1	-1	1193	807	-1	-1	1195	-1	
999	7	1	810	1194	1195	808	1196	807	1190	
1000	7	2	1194	826	-1	817	1197	809	1189	810(2)
1001	7	1	1196	1197	-1	1180	810	-1	1191	
1002	7	2	1198	812	811	1199	-1	-1	823	
1003	7	2	812	1198	822	1187	-1	-1	824	814(3)
1004	7	1	813	814	1198	1200	-1	-1	825	
1005	7	1	1200	1189	-1	813	1201	-1	826	
1006	7	2	1202	816	815	1203	1192	827	1186	
1007	7	2	816	1202	818	824	1183	820	1187	
1008	7	2	817	818	1202	1194	1182	819	1188	
1009	7	2	827	820	819	1204	827	1202	-1	
1010	7	1	821	822	1198	1205	-1	-1	828	
1011	7	1	1205	-1	-1	821	1206	-1	829	
1012	7	2	823	824	828	1207	-1	830	1198	825(3)
1013	7	2	1207	1203	-1	823	1208	831	1199	
1014	7	1	826	1194	1209	825	1210	832	1200	
1015	7	1	1210	1197	-1	-1	826	-1	1201	
1016	7	2	1204	831	-1	827	1211	1203	-1	
1017	7	1	829	-1	1212	828	1213	834	1205	
1018	7	1	1213	-1	-1	-1	829	1214	1206	
1019	7	2	831	1204	1215	830	1216	1207	-1	
1020	7	2	1216	1211	-1	836	831	1217	-1	
1021	7	1	-1	1193	832	1215	-1	1209	-1	
1022	7	2	836	836	837	1216	833	1218	-1	
1023	7	1	-1	-1	834	1215	1219	1212	-1	
1024	7	2	835	-1	1219	837	841	1220	-1	834(5)
1025	7	2	1220	-1	-1	838	1214	835	1221	
1026	7	2	1218	1218	839	1217	-1	836	1222	
1027	7	2	838	839	1218	1220	-1	837	1223	
1028	7	2	1223	840	840	1221	-1	-1	838	
1029	7	2	840	1223	1222	-1	-1	-1	839	
1030	7	1	-1	-1	841	-1	1219	1224	-1	
1031	7	1	844	-1	1224	842	1225	841	1226	
1032	7	1	845	843	-1	1226	-1	-1	842	
1033	7	1	1225	-1	-1	-1	844	1214	1227	
1034	7	1	1226	-1	-1	845	1227	-1	844	
1035	8	1	847	-1	1228	846	1229	-1	848	
1036	8	1	1229	-1	-1	-1	847	1230	849	
1037	8	1	-1	-1	848	-1	1231	850	1228	
1038	8	1	849	-1	1231	-1	848	1232	1229	
1039	8	2	1232	-1	-1	-1	852	867	1233	849(6)
1040	8	2	-1	-1	851	1234	861	1235	-1	850(5)
1041	8	2	852	-1	1235	854	1232	851	1236	
1042	8	2	1236	-1	-1	855	1233	-1	852	
1043	8	2	856	1237	1234	853	859	1238	-1	
1044	8	2	857	857	854	1238	-1	853	1239	
1045	8	2	855	858	1239	1236	-1	-1	854	
1046	8	2	1237	856	-1	1237	869	1240	-1	
1047	8	2	1238	1240	-1	857	1241	856	1242	
1048	8	2	1239	1239	858	1242	-1	-1	857	
1049	8	2	869	859	1243	864	1237	1244	-1	
1050	8	2	-1	1243	859	876	1234	1245	-1	861(4)
1051	8	2	870	1244	1245	860	1246	859	1247	
1052	8	2	871	866	878	1247	-1	-1	860	863(3)
1053	8	1	-1	-1	862	1245	1248	861	1249	
1054	8	1	867	-1	1248	-1	862	1232	1250	
1055	8	1	868	-1	1249	863	1250	-1	862	
1056	8	2	872	865	882	1244	1251	864	1252	

Parameter	ℓ	d	Simples 1	2	3	4	5	6	7	Other
1057	8	2	873	1251	883	-1	865	873	1253	
1058	8	2	874	1252	884	866	1253	-1	865	
1059	8	1	1250	-1	-1	-1	868	1254	867	
1060	8	2	1244	870	-1	872	1255	869	1256	
1061	8	2	1246	1255	-1	-1	870	1241	1257	
1062	8	2	1247	1256	-1	871	1257	-1	870	
1063	8	2	1251	873	888	1258	872	1251	1259	
1064	8	2	1252	874	889	1256	1259	-1	872	
1065	8	2	1253	1259	890	-1	874	1260	873	
1066	8	2	-1	894	875	1243	-1	1261	-1	876(2)
1067	8	2	879	895	1261	882	1262	875	1263	877(2)
1068	8	1	-1	1261	877	1245	1264	876	1265	
1069	8	1	880	1262	1264	-1	877	-1	1266	
1070	8	1	881	1263	1265	878	1266	-1	877	
1071	8	2	1262	902	-1	1267	879	-1	1268	880(2)
1072	8	2	1263	903	-1	886	1268	-1	879	881(2)
1073	8	1	1266	1268	-1	-1	881	1269	880	
1074	8	2	885	883	1251	1270	882	888	1271	
1075	8	2	886	884	1252	1263	1271	-1	882	
1076	8	2	887	1271	1253	900	884	1272	883	
1077	8	2	1270	904	1273	885	1267	891	1274	
1078	8	2	1271	887	892	1274	886	1275	885	
1079	8	2	1272	1275	893	909	-1	887	893	
1080	8	2	891	898	1276	888	-1	1270	1277	
1081	8	2	892	890	1259	1277	889	1278	888	
1082	8	2	893	1278	1260	901	-1	890	1272	
1083	8	2	1279	910	891	1273	-1	1273	1280	
1084	8	2	1277	911	1280	892	-1	1281	891	
1085	8	2	1278	893	1275	1281	-1	892	1275	
1086	8	1	-1	1261	895	-1	1282	894	1283	
1087	8	1	902	1262	1282	1284	895	-1	1285	
1088	8	1	903	1263	1283	897	1285	-1	895	
1089	8	2	904	1270	1286	896	1284	898	1287	
1090	8	2	905	900	-1	1287	916	1288	896	897(5)
1091	8	2	1289	1276	898	-1	-1	1286	1290	
1092	8	2	911	1277	1290	899	-1	1291	898	
1093	8	2	912	901	-1	1291	917	899	1288	
1094	8	2	909	1288	-1	1272	918	900	901	
1095	8	2	1284	1267	1292	925	904	-1	1293	902(4)
1096	8	1	1285	1268	-1	1293	903	1294	902	
1097	8	2	1295	1273	904	906	1292	910	1296	
1098	8	2	1287	1274	1296	905	1293	1297	904	
1099	8	2	1288	909	908	1297	928	905	912	
1100	8	2	906	919	-1	1295	932	919	1298	
1101	8	2	1298	920	907	1296	933	1299	906	
1102	8	2	-1	921	1299	908	1300	907	921	
1103	8	2	910	1279	1289	919	-1	1295	1301	
1104	8	2	1301	1280	911	920	-1	1302	910	
1105	8	2	1291	1281	1302	912	1303	911	1297	
1106	8	2	959	-1	922	-1	-1	1304	1305	913(1)
1107	8	1	1304	-1	914	-1	1306	913	1307	
1108	8	2	1305	-1	915	-1	-1	1308	929	913(7)
1109	8	1	925	-1	1306	1284	914	-1	1309	
1110	8	1	926	-1	1307	916	1309	1310	914	
1111	8	2	927	-1	1308	917	1311	915	1310	
1112	8	2	928	918	-1	1310	1288	931	917	916(6)
1113	8	2	920	1298	-1	1301	935	1312	919	
1114	8	2	1312	1299	921	1302	1313	920	1299	
1115	8	3	-1	1300	1313	940	964	937	1300	921(5)
1116	8	2	923	-1	1304	932	1314	922	1315	
1117	8	3	924	-1	1305	938	-1	1316	941	922(7)
1118	8	2	1314	-1	947	1292	923	-1	1317	925(3)
1119	8	2	1315	-1	948	933	1317	1318	923	926(3)
1120	8	3	1316	-1	962	939	1319	924	1318	927(3)
1121	8	1	1309	-1	1317	1293	926	1320	925	
1122	8	2	1310	-1	1318	928	1321	944	927	926(6)
1123	8	2	1311	-1	1319	1303	927	-1	1321	
1124	8	2	980	-1	941	-1	-1	1322	1305	929(1)
1125	8	1	1322	-1	930	-1	1323	929	1324	
1126	8	1	943	-1	1323	-1	930	-1	1325	
1127	8	1	944	-1	1324	931	1325	1310	930	
1128	8	2	933	935	-1	1315	1298	1326	932	
1129	8	3	1326	936	934	1318	1327	957	939	933(6)
1130	8	3	-1	937	1327	1300	934	1300	940	
1131	8	3	936	1326	-1	-1	1328	970	953	935(6)
1132	8	3	1328	1327	937	1329	936	1313	954	
1133	8	3	939	953	-1	1316	1330	938	1326	
1134	8	3	1330	954	940	1331	939	954	1327	
1135	8	2	942	-1	1322	955	1332	941	1333	
1136	8	2	1332	-1	983	1334	942	-1	1335	943(3)
1137	8	2	1333	-1	979	957	1335	1318	942	944(3)
1138	8	1	1325	-1	1335	-1	944	1336	943	
1139	8	1	1304	-1	945	-1	1337	959	1338	
1140	8	1	947	-1	1337	-1	945	-1	1339	
1141	8	1	948	-1	1338	946	1339	1340	945	
1142	8	2	949	951	-1	1340	1341	978	961	946(6)
1143	8	1	1339	-1	1317	-1	948	1342	947	
1144	8	2	1340	-1	1318	949	1343	979	962	948(6)
1145	8	2	1341	952	950	1344	949	976	963	
1146	8	2	-1	964	1345	950	1300	958	964	
1147	8	2	952	1341	-1	1346	951	977	965	
1148	8	2	1346	1347	966	952	-1	975	966	
1149	8	3	954	1330	-1	1348	953	1330	1328	
1150	8	3	1348	1349	1329	954	-1	1331	1329	
1151	8	2	956	971	-1	1350	955	971	1351	
1152	8	2	957	970	-1	1333	1351	1326	955	
1153	8	2	1350	1352	984	956	1334	972	1353	
1154	8	2	1351	969	958	1353	957	1345	956	
1155	8	2	1305	-1	960	-1	-1	1354	980	959(7)
1156	8	2	962	-1	1354	961	1355	960	1340	
1157	8	2	963	965	-1	1356	961	967	1341	
1158	8	2	1355	-1	1319	1357	962	-1	1343	
1159	8	2	1356	1358	1359	963	1357	968	1344	
1160	8	2	1360	1345	964	1359	1313	969	1345	
1161	8	2	966	1361	1362	965	-1	973	1346	
1162	8	2	1361	966	1347	1358	-1	974	1347	
1163	8	2	968	1363	1364	967	-1	1356	987	
1164	8	2	1363	968	1365	974	-1	1358	988	
1165	8	2	1366	1365	968	969	-1	1359	972	
1166	8	2	969	1351	-1	1366	970	1360	971	
1167	8	2	972	1367	1368	971	-1	1350	1366	
1168	8	2	1367	972	988	1352	-1	1352	1365	
1169	8	2	974	973	1369	1363	989	1361	995	
1170	8	2	1370	1369	973	-1	990	1362	996	
1171	8	2	1371	975	974	1372	991	1347	986	
1172	8	2	975	1371	1370	977	992	1346	985	
1173	8	2	976	977	-1	1373	993	1341	982	978(5)
1174	8	2	1373	1372	1353	976	1374	1344	984	
1175	8	1	979	-1	1375	978	1376	1340	981	
1176	8	1	1376	-1	1335	1374	979	1377	983	
1177	8	1	1322	-1	981	-1	1378	980	1375	
1178	8	1	983	-1	1378	1379	981	-1	1376	
1179	8	2	984	1380	1381	982	1379	987	1373	
1180	8	2	1379	1382	1334	1001	984	-1	1374	983(4)
1181	8	2	1380	984	1352	986	1382	988	1372	
1182	8	2	986	985	1383	1380	1008	995	1371	
1183	8	2	996	1383	985	-1	1007	996	1370	
1184	8	2	988	987	1384	995	-1	1380	1363	
1185	8	2	1368	1384	987	-1	-1	1381	1364	
1186	8	2	1385	990	989	1386	1369	-1	1006	
1187	8	2	990	1385	992	1003	1370	-1	1007	
1188	8	2	991	992	1385	1387	1371	-1	1008	
1189	8	2	1387	1005	-1	991	1388	-1	1000	994(2)
1190	8	1	994	1387	1389	993	1390	-1	999	
1191	8	1	1390	1388	-1	1374	994	1391	1001	
1192	8	2	1383	996	995	1392	1006	1383	1369	
1193	8	2	-1	1021	997	1393	-1	1394	-1	998(2)
1194	8	2	1000	1014	1394	1008	1395	997	1387	999(2)
1195	8	1	-1	1394	999	-1	1396	998	1389	
1196	8	1	1001	1395	1396	1379	999	-1	1390	
1197	8	2	1395	1015	-1	1382	1000	-1	1388	1001(2)
1198	8	2	1002	1003	1010	1397	-1	-1	1012	1004(3)
1199	8	2	1397	1386	-1	1002	1398	-1	1013	
1200	8	1	1005	1387	1399	1004	1400	-1	1014	
1201	8	1	1400	1388	-1	-1	1005	1401	1015	
1202	8	2	1006	1007	1008	1402	1383	1009	1385	
1203	8	2	1402	1013	-1	1006	1403	1016	1386	
1204	8	2	1016	1019	1393	1009	1404	1402	-1	
1205	8	1	1011	-1	1405	1010	1406	-1	1017	
1206	8	1	1406	-1	-1	-1	1011	1407	1018	
1207	8	2	1013	1402	1408	1012	1409	1019	1397	
1208	8	2	1409	1403	-1	-1	1013	1410	1398	

Parameter	ℓ	d	Simples 1	2	3	4	5	6	7	Other
1209	8	1	-1	1394	1014	1408	1411	1021	1399	
1210	8	1	1015	1395	1411	-1	1014	-1	1400	
1211	8	2	1404	1020	-1	1404	1016	1412	-1	
1212	8	1	-1	-1	1017	1408	1413	1023	1405	
1213	8	1	1018	-1	1413	-1	1017	1414	1406	
1214	8	2	1414	-1	-1	-1	1025	1033	1415	1018(6)
1215	8	2	-1	1393	1019	1023	1416	1408	-1	1021(4)
1216	8	2	1020	1404	1416	1022	1019	1417	-1	
1217	8	2	1417	1412	-1	1026	1410	1020	1418	
1218	8	2	1026	1026	1027	1417	-1	1022	1419	
1219	8	2	-1	-1	1024	1416	1030	1420	-1	1023(5)
1220	8	2	1025	-1	1420	1027	1414	1024	1421	
1221	8	2	1421	-1	-1	1028	1415	-1	1025	
1222	8	2	1419	1419	1029	1418	-1	-1	1026	
1223	8	2	1028	1029	1419	1421	-1	-1	1027	
1224	8	1	-1	-1	1031	-1	1422	1030	1423	
1225	8	1	1033	-1	1422	-1	1031	1414	1424	
1226	8	1	1034	-1	1423	1032	1424	-1	1031	
1227	8	1	1424	-1	-1	-1	1034	1425	1033	
1228	9	1	-1	-1	1035	-1	1426	-1	1037	
1229	9	1	1036	-1	1426	-1	1035	1427	1038	
1230	9	1	1427	-1	-1	-1	-1	1036	1428	
1231	9	1	-1	-1	1038	1429	1037	1430	1426	
1232	9	2	1039	-1	1430	-1	1041	1054	1431	1038(6)
1233	9	2	1431	-1	-1	-1	1042	1428	1039	
1234	9	2	-1	1432	1043	1040	1050	1433	-1	
1235	9	2	-1	-1	1041	1433	1430	1040	1434	
1236	9	2	1042	-1	1434	1045	1431	-1	1041	
1237	9	2	1046	1043	1432	1046	1049	1435	-1	
1238	9	2	1047	1435	1433	1044	1436	1043	1437	
1239	9	2	1048	1048	1045	1437	-1	-1	1044	
1240	9	2	1435	1047	-1	1435	1438	1046	1439	
1241	9	2	1436	1438	-1	-1	1047	1061	1440	
1242	9	2	1437	1439	-1	1048	1440	-1	1047	
1243	9	2	-1	1050	1049	1066	1432	1441	-1	
1244	9	2	1060	1051	1441	1056	1442	1049	1443	
1245	9	2	-1	1441	1051	1068	1444	1050	1445	1053(4)
1246	9	2	1061	1442	1444	-1	1051	1436	1446	
1247	9	2	1062	1443	1445	1052	1446	-1	1051	
1248	9	1	-1	-1	1054	1447	1053	1430	1448	
1249	9	1	-1	-1	1055	1445	1448	-1	1053	
1250	9	1	1059	-1	1448	-1	1055	1449	1054	
1251	9	2	1063	1057	1074	1450	1056	1063	1451	
1252	9	2	1064	1058	1075	1443	1451	-1	1056	
1253	9	2	1065	1451	1076	-1	1058	1452	1057	
1254	9	1	1449	-1	-1	-1	-1	1059	1428	
1255	9	2	1442	1061	-1	1453	1060	1438	1454	
1256	9	2	1443	1062	-1	1064	1454	-1	1060	
1257	9	2	1446	1454	-1	-1	1062	1455	1061	
1258	9	2	1450	-1	1456	1063	1453	1450	1457	
1259	9	2	1451	1065	1081	1457	1064	1458	1063	
1260	9	2	1452	1458	1082	-1	-1	1065	1452	
1261	9	2	-1	1086	1067	1441	1459	1066	1460	1068(2)
1262	9	2	1071	1087	1459	1461	1067	-1	1462	1069(2)
1263	9	2	1072	1088	1460	1075	1462	-1	1067	1070(2)
1264	9	1	-1	1459	1069	1463	1068	-1	1464	
1265	9	1	-1	1460	1070	1445	1464	-1	1068	
1266	9	1	1073	1462	1464	-1	1070	1465	1069	
1267	9	2	1461	1095	1466	1071	1077	-1	1467	
1268	9	2	1462	1096	-1	1467	1072	1468	1071	1073(2)
1269	9	1	1465	1468	-1	-1	-1	1073	-1	
1270	9	2	1077	1089	1469	1074	1461	1080	1470	
1271	9	2	1078	1076	1451	1470	1075	1471	1074	
1272	9	2	1079	1471	1452	1094	-1	1076	1082	
1273	9	2	1472	1097	1077	1083	1466	1083	1473	
1274	9	2	1470	1098	1473	1078	1467	1474	1077	
1275	9	2	1471	1079	1085	1474	-1	1078	1085	
1276	9	2	1475	1091	1080	1456	-1	1469	1476	
1277	9	2	1084	1092	1476	1081	-1	1477	1080	
1278	9	2	1085	1082	1458	1477	-1	1081	1471	
1279	9	2	1083	1103	1475	1472	-1	1472	1478	
1280	9	2	1478	1104	1084	1473	-1	1479	1083	
1281	9	2	1477	1105	1479	1085	1480	1084	1474	
1282	9	1	-1	1459	1087	1481	1086	-1	1482	
1283	9	1	-1	1460	1088	-1	1482	-1	1086	
1284	9	2	1095	1461	1483	1109	1089	-1	1484	1087(4)

Parameter	ℓ	d	Simples 1	2	3	4	5	6	7	Other
1285	9	1	1096	1462	1482	1484	1088	1485	1087	
1286	9	2	1486	1469	1089	-1	1483	1091	1487	
1287	9	2	1098	1470	1487	1090	1484	1488	1089	
1288	9	2	1099	1094	-1	1488	1112	1090	1093	
1289	9	2	1091	1475	1103	-1	-1	1486	1489	
1290	9	2	1489	1476	1092	-1	-1	1490	1091	
1291	9	2	1105	1477	1490	1093	1491	1092	1488	
1292	9	2	1492	1466	1095	1118	1097	-1	1493	
1293	9	2	1484	1467	1493	1121	1098	1494	1095	1096(4)
1294	9	1	1485	1468	-1	1494	-1	1096	-1	
1295	9	2	1097	1472	1486	1100	1492	1103	1495	
1296	9	2	1495	1473	1098	1101	1493	1496	1097	
1297	9	2	1488	1474	1496	1099	1497	1098	1105	
1298	9	2	1101	1113	-1	1495	1128	1498	1100	
1299	9	2	1498	1114	1102	1496	1499	1101	1114	
1300	9	3	-1	1115	1499	1130	1146	1130	1115	1102(5)
1301	9	2	1104	1478	1489	1113	-1	1500	1103	
1302	9	2	1500	1479	1105	1114	1501	1104	1496	
1303	9	2	1491	1480	1501	1123	1105	-1	1497	
1304	9	2	1139	-1	1116	-1	1502	1106	1503	1107(1)
1305	9	3	1155	-1	1117	-1	-1	1504	1124	1108(1) 1106(7)
1306	9	1	1502	-1	1109	1505	1107	-1	1506	
1307	9	1	1503	-1	1110	-1	1506	1507	1107	
1308	9	2	1504	-1	1111	-1	1508	1108	1507	
1309	9	1	1121	-1	1506	1484	1110	1509	1109	
1310	9	2	1122	-1	1507	1112	1510	1127	1111	1110(6)
1311	9	2	1123	-1	1508	1491	1111	-1	1510	
1312	9	2	1114	1498	-1	1500	1511	1113	1498	
1313	9	3	1511	1499	1115	1512	1160	1132	1499	1114(5)
1314	9	2	1118	-1	1502	1492	1116	-1	1513	
1315	9	2	1119	-1	1503	1128	1513	1514	1116	
1316	9	3	1120	-1	1504	1133	1515	1117	1514	
1317	9	2	1513	-1	1143	1493	1119	1516	1118	1121(3)
1318	9	3	1514	-1	1144	1129	1517	1137	1120	1122(3) 1119(6)
1319	9	3	1515	-1	1158	1518	1120	-1	1517	1123(3)
1320	9	1	1509	-1	1516	1494	1519	1121	-1	
1321	9	2	1510	-1	1517	1497	1122	1519	1123	
1322	9	2	1177	-1	1135	-1	1520	1124	1521	1125(1)
1323	9	1	1520	-1	1126	1522	1125	-1	1523	
1324	9	1	1521	-1	1127	-1	1523	1507	1125	
1325	9	1	1138	-1	1523	-1	1127	1524	1126	
1326	9	3	1129	1131	-1	1514	1525	1152	1133	1128(6)
1327	9	3	1525	1132	1130	1526	1129	1499	1134	
1328	9	3	1132	1525	-1	1527	1131	1511	1149	
1329	9	3	1527	1528	1150	1132	-1	1512	1150	
1330	9	3	1134	1149	-1	1529	1133	1149	1525	
1331	9	3	1529	1530	1512	1134	1518	1150	1526	
1332	9	2	1136	-1	1520	1531	1135	-1	1532	
1333	9	2	1137	-1	1521	1152	1532	1514	1135	
1334	9	2	1531	1533	1180	1136	1153	-1	1534	
1335	9	2	1532	-1	1176	1534	1137	1535	1136	1138(3)
1336	9	1	1524	-1	1535	-1	1519	1138	-1	
1337	9	1	1502	-1	1140	1536	1139	-1	1537	
1338	9	1	1503	-1	1141	-1	1537	1538	1139	
1339	9	1	1143	-1	1537	-1	1141	1539	1140	
1340	9	2	1144	-1	1538	1142	1540	1175	1156	1141(6)
1341	9	2	1145	1147	-1	1541	1142	1173	1157	
1342	9	1	1539	-1	1516	-1	1542	1143	-1	
1343	9	2	1540	-1	1517	1543	1144	1542	1158	
1344	9	2	1541	1544	1545	1145	1543	1174	1159	
1345	9	2	1546	1160	1146	1545	1499	1154	1160	
1346	9	2	1148	1547	1548	1147	-1	1172	1161	
1347	9	2	1547	1148	1162	1544	-1	1171	1162	
1348	9	3	1150	1549	1550	1149	-1	1529	1527	
1349	9	3	1549	1150	1528	1530	-1	1530	1528	
1350	9	2	1153	1551	1552	1151	1531	1167	1553	
1351	9	2	1154	1166	-1	1553	1152	1546	1151	
1352	9	2	1551	1153	1181	1168	1533	1168	1554	
1353	9	2	1553	1554	1174	1154	1534	1545	1153	
1354	9	2	1504	-1	1156	-1	1555	1155	1538	
1355	9	2	1158	-1	1555	1556	1156	-1	1540	
1356	9	2	1159	1557	1558	1157	1556	1163	1541	
1357	9	2	1556	1559	1560	1158	1159	-1	1543	
1358	9	2	1557	1159	1561	1162	1559	1164	1544	
1359	9	2	1562	1561	1159	1160	1560	1165	1545	
1360	9	2	1160	1546	-1	1562	1511	1166	1546	

Para-meter	ℓ	d	Simples 1	2	3	4	5	6	7	Other
1361	9	2	1162	1161	1563	1557	-1	1169	1547	
1362	9	2	1548	1563	1161	-1	-1	1170	1548	
1363	9	2	1164	1163	1564	1169	-1	1557	1184	
1364	9	2	1565	1564	1163	-1	-1	1558	1185	
1365	9	2	1566	1165	1164	1554	-1	1561	1168	
1366	9	2	1165	1566	1565	1166	-1	1562	1167	
1367	9	2	1168	1167	1567	1551	-1	1551	1566	
1368	9	2	1185	1567	1167	-1	-1	1552	1565	
1369	9	2	1568	1170	1169	1569	1186	1563	1192	
1370	9	2	1170	1568	1172	-1	1187	1548	1183	
1371	9	2	1171	1172	1568	1570	1188	1547	1182	
1372	9	2	1570	1174	1554	1171	1571	1544	1181	
1373	9	2	1174	1570	1572	1173	1573	1541	1179	
1374	9	2	1573	1571	1534	1191	1174	1574	1180	1176(4)
1375	9	1	1521	-1	1175	-1	1575	1538	1177	
1376	9	1	1176	-1	1575	1573	1175	1576	1178	
1377	9	1	1576	-1	1535	1574	1542	1176	-1	
1378	9	1	1520	-1	1178	1577	1177	-1	1575	
1379	9	2	1180	1578	1579	1196	1179	-1	1573	1178(4)
1380	9	2	1181	1179	1580	1182	1578	1184	1570	
1381	9	2	1552	1580	1179	-1	1579	1185	1572	
1382	9	2	1578	1180	1533	1197	1181	-1	1571	
1383	9	2	1192	1183	1182	1581	1202	1192	1568	
1384	9	2	1567	1185	1184	1582	-1	1580	1564	
1385	9	2	1186	1187	1188	1583	1568	-1	1202	
1386	9	2	1583	1199	-1	1186	1584	-1	1203	
1387	9	2	1189	1200	1585	1188	1586	-1	1194	1190(2)
1388	9	2	1586	1201	-1	1571	1189	1587	1197	1191(2)
1389	9	1	-1	1585	1190	-1	1588	-1	1195	
1390	9	1	1191	1586	1588	1573	1190	1589	1196	
1391	9	1	1589	1587	-1	1574	-1	1191	-1	
1392	9	2	1581	-1	1582	1192	1590	1581	1569	
1393	9	2	-1	1215	1204	1193	1591	1592	-1	
1394	9	2	-1	1209	1194	1592	1593	1193	1585	1195(2)
1395	9	2	1197	1210	1593	1578	1194	-1	1586	1196(2)
1396	9	1	-1	1593	1196	1594	1195	-1	1588	
1397	9	2	1199	1583	1595	1198	1596	-1	1207	
1398	9	2	1596	1584	-1	-1	1199	1597	1208	
1399	9	1	-1	1585	1200	1595	1598	-1	1209	
1400	9	1	1201	1586	1598	-1	1200	1599	1210	
1401	9	1	1599	1587	-1	-1	-1	1201	-1	
1402	9	2	1203	1207	1592	1202	1600	1204	1583	
1403	9	2	1600	1208	-1	1590	1203	1601	1584	
1404	9	2	1211	1216	1591	1211	1204	1602	-1	
1405	9	1	-1	-1	1205	1595	1603	-1	1212	
1406	9	1	1206	-1	1603	-1	1205	1604	1213	
1407	9	1	1604	-1	-1	-1	-1	1206	1605	
1408	9	2	-1	1592	1207	1212	1606	1215	1595	1209(4)
1409	9	2	1208	1600	1606	-1	1207	1607	1596	
1410	9	2	1607	1601	-1	-1	1217	1208	1608	
1411	9	1	-1	1593	1210	1609	1209	-1	1598	
1412	9	2	1602	1217	-1	1602	1601	1211	1610	
1413	9	1	-1	-1	1213	1611	1212	1612	1603	
1414	9	2	1214	-1	1612	-1	1220	1225	1613	1213(6)
1415	9	2	1613	-1	-1	-1	1221	1605	1214	
1416	9	2	-1	1591	1216	1219	1215	1614	-1	
1417	9	2	1217	1602	1614	1218	1607	1216	1615	
1418	9	2	1615	1610	-1	1222	1608	-1	1217	
1419	9	2	1222	1222	1223	1615	-1	-1	1218	
1420	9	2	-1	-1	1220	1614	1612	1219	1616	
1421	9	2	1221	-1	1616	1223	1613	-1	1220	
1422	9	1	-1	-1	1225	1617	1224	1612	1618	
1423	9	1	-1	-1	1226	-1	1618	-1	1224	
1424	9	1	1227	-1	1618	-1	1226	1619	1225	
1425	9	1	1619	-1	-1	-1	-1	1227	1605	
1426	10	1	-1	-1	1229	1620	1228	1621	1231	
1427	10	1	1230	-1	1621	-1	-1	1229	1622	
1428	10	2	1622	-1	-1	-1	-1	1233	1254	1230(7)
1429	10	1	-1	1623	-1	1231	-1	1624	1620	
1430	10	2	-1	-1	1232	1624	1235	1248	1625	1231(6)
1431	10	2	1233	-1	1625	-1	1236	1622	1232	
1432	10	2	-1	1234	1237	-1	1243	1626	-1	
1433	10	2	-1	1626	1238	1235	1627	1234	1628	
1434	10	2	-1	-1	1236	1628	1625	-1	1235	
1435	10	2	1240	1238	1626	1240	1629	1237	1630	
1436	10	2	1241	1629	1627	-1	1238	1246	1631	
1437	10	2	1242	1630	1628	1239	1631	-1	1238	
1438	10	2	1629	1241	-1	1632	1240	1255	1633	
1439	10	2	1630	1242	-1	1630	1633	-1	1240	
1440	10	2	1631	1633	-1	-1	1242	1634	1241	
1441	10	2	-1	1245	1244	1261	1635	1243	1636	
1442	10	2	1255	1246	1635	1637	1244	1629	1638	
1443	10	2	1256	1247	1636	1252	1638	-1	1244	
1444	10	2	-1	1635	1246	1639	1245	1627	1640	
1445	10	2	-1	1636	1247	1265	1640	-1	1245	1249(4)
1446	10	2	1257	1638	1640	-1	1247	1641	1246	
1447	10	1	-1	1642	-1	1248	1639	1624	1643	
1448	10	1	-1	-1	1250	1643	1249	1644	1248	
1449	10	1	1254	-1	1644	-1	-1	1250	1622	
1450	10	2	1258	-1	1645	1251	1637	1258	1646	
1451	10	2	1259	1253	1271	1646	1252	1647	1251	
1452	10	2	1260	1647	1272	-1	-1	1253	1260	
1453	10	2	1637	-1	1648	1255	1258	1632	1649	
1454	10	2	1638	1257	-1	1649	1256	1650	1255	
1455	10	2	1641	1650	-1	-1	-1	1257	1634	
1456	10	2	1651	-1	1258	1276	1648	1645	1652	
1457	10	2	1646	-1	1652	1259	1649	1653	1258	
1458	10	2	1647	1260	1278	1653	-1	1259	1647	
1459	10	2	-1	1282	1262	1654	1261	-1	1655	1264(2)
1460	10	2	-1	1283	1263	1636	1655	-1	1261	1265(2)
1461	10	2	1267	1284	1656	1262	1270	-1	1657	
1462	10	2	1268	1285	1655	1657	1263	1658	1262	1266(2)
1463	10	1	-1	1659	-1	1264	1639	-1	1660	
1464	10	1	-1	1655	1266	1660	1265	1661	1264	
1465	10	1	1269	1658	1661	-1	-1	1266	-1	
1466	10	2	1662	1292	1267	-1	1273	-1	1663	
1467	10	2	1657	1293	1663	1268	1274	1664	1267	
1468	10	2	1658	1294	-1	1664	-1	1268	-1	1269(2)
1469	10	2	1665	1286	1270	1645	1656	1276	1666	
1470	10	2	1274	1287	1666	1271	1657	1667	1270	
1471	10	2	1275	1272	1647	1667	-1	1271	1278	
1472	10	2	1273	1295	1665	1279	1662	1279	1668	
1473	10	2	1668	1296	1274	1280	1663	1669	1273	
1474	10	2	1667	1297	1669	1275	1670	1274	1281	
1475	10	2	1276	1289	1279	1651	-1	1665	167.	
1476	10	2	1671	1290	1277	1652	-1	1672	1276	
1477	10	2	1281	1291	1672	1278	1673	1277	1667	
1478	10	2	1280	1301	1671	1668	-1	1674	1279	
1479	10	2	1674	1302	1281	1669	1675	1280	1669	
1480	10	2	1673	1303	1675	-1	1281	-1	1670	
1481	10	1	-1	1676	1677	1282	-1	-1	1678	
1482	10	1	-1	1655	1285	1678	1283	1679	1282	
1483	10	2	1680	1656	1284	1677	1286	-1	1681	
1484	10	2	1293	1657	1681	1309	1287	1682	1284	1285(4)
1485	10	1	1294	1658	1679	1682	-1	1285	-1	
1486	10	2	1286	1665	1295	-1	1680	1289	1683	
1487	10	2	1683	1666	1287	-1	1681	1684	1286	
1488	10	2	1297	1667	1684	1288	1685	1287	1291	
1489	10	2	1290	1671	1301	-1	-1	1686	1289	
1490	10	2	1686	1672	1291	-1	1687	1290	1684	
1491	10	2	1303	1673	1687	1311	1291	-1	1685	
1492	10	2	1292	1662	1680	1314	1295	-1	1688	
1493	10	2	1688	1663	1293	1317	1296	1689	1292	
1494	10	2	1682	1664	1689	1320	1690	1293	-1	1294(4)
1495	10	2	1296	1668	1683	1298	1688	1691	1295	
1496	10	2	1691	1669	1297	1299	1692	1296	1302	
1497	10	2	1685	1670	1692	1321	1297	1690	1303	
1498	10	2	1299	1312	-1	1691	1693	1298	1312	
1499	10	3	1693	1313	1300	1694	1345	1327	1313	1299(5)
1500	10	2	1302	1674	1686	1312	1695	1301	1691	
1501	10	2	1695	1675	1303	1696	1302	-1	1692	
1502	10	2	1337	-1	1314	1697	1304	-1	1698	1306(1)
1503	10	2	1338	-1	1315	-1	1698	1699	1304	1307(1)
1504	10	3	1354	-1	1316	-1	1700	1305	1699	1308(1)
1505	10	1	1697	1701	1677	1306	-1	-1	1702	
1506	10	1	1698	-1	1309	1702	1307	1703	1306	
1507	10	2	1699	-1	1310	-1	1704	1324	1308	1307(6)
1508	10	2	1700	-1	1311	1705	1308	-1	1704	
1509	10	1	1320	-1	1703	1682	1706	1309	-1	
1510	10	2	1321	-1	1704	1685	1310	1706	1311	
1511	10	3	1313	1693	-1	1707	1360	1328	1693	1312(5)
1512	10	3	1707	1708	1331	1313	1696	1329	1694	

Parameter	ℓ	d	Simples							Other
			1	2	3	4	5	6	7	
1513	10	2	1317	-1	1698	1688	1315	1709	1314	
1514	10	3	1318	-1	1699	1326	1710	1333	1316	1315(6)
1515	10	3	1319	-1	1700	1711	1316	-1	1710	
1516	10	2	1709	-1	1342	1689	1712	1317	-1	1320(3)
1517	10	3	1710	-1	1343	1713	1318	1712	1319	1321(3)
1518	10	3	1711	1714	1696	1319	1331	-1	1713	
1519	10	2	1706	-1	1712	1690	1336	1321	-1	1320(5)
1520	10	2	1378	-1	1332	1715	1322	-1	1716	1323(1)
1521	10	2	1375	-1	1333	-1	1716	1699	1322	1324(1)
1522	10	1	1715	1717	-1	1323	-1	-1	1718	
1523	10	1	1716	-1	1325	1718	1324	1719	1323	
1524	10	1	1336	-1	1719	-1	1706	1325	-1	
1525	10	3	1327	1328	-1	1720	1326	1693	1330	
1526	10	3	1720	1721	1694	1327	1713	1694	1331	
1527	10	3	1329	1722	1723	1328	-1	1707	1348	
1528	10	3	1722	1329	1349	1721	-1	1708	1349	
1529	10	3	1331	1724	1725	1330	1711	1348	1720	
1530	10	3	1724	1331	1708	1349	1714	1349	1721	
1531	10	2	1334	1726	1727	1332	1350	-1	1728	
1532	10	2	1335	-1	1716	1728	1333	1729	1332	
1533	10	2	1726	1334	1382	-1	1352	-1	1730	
1534	10	2	1728	1730	1374	1335	1353	1731	1334	
1535	10	2	1729	-1	1377	1731	1712	1335	-1	1336(3)
1536	10	1	1697	1732	-1	1337	-1	-1	1733	
1537	10	1	1698	-1	1339	1733	1338	1734	1337	
1538	10	2	1699	-1	1340	-1	1735	1375	1354	1338(6)
1539	10	1	1342	-1	1734	-1	1736	1339	-1	
1540	10	2	1343	-1	1735	1737	1340	1736	1355	
1541	10	2	1344	1738	1739	1341	1737	1373	1356	
1542	10	2	1736	-1	1712	1740	1377	1343	-1	1342(5)
1543	10	2	1737	1741	1742	1343	1344	1740	1357	
1544	10	2	1738	1344	1743	1347	1741	1372	1358	
1545	10	2	1744	1743	1344	1345	1742	1353	1359	
1546	10	2	1345	1360	-1	1744	1693	1351	1360	
1547	10	2	1347	1346	1745	1738	-1	1371	1361	
1548	10	2	1362	1745	1346	-1	-1	1370	1362	
1549	10	3	1349	1348	1746	1724	-1	1724	1722	
1550	10	3	1723	1746	1348	-1	-1	1725	1723	
1551	10	2	1352	1350	1747	1367	1726	1367	1748	
1552	10	2	1381	1747	1350	-1	1727	1368	1749	
1553	10	2	1353	1748	1749	1351	1728	1744	1350	
1554	10	2	1748	1353	1372	1365	1730	1743	1352	
1555	10	2	1700	-1	1355	1750	1354	-1	1735	
1556	10	2	1357	1751	1752	1355	1356	-1	1737	
1557	10	2	1358	1356	1753	1361	1751	1363	1738	
1558	10	2	1754	1753	1356	-1	1752	1364	1739	
1559	10	2	1751	1357	1755	-1	1358	-1	1741	
1560	10	2	1756	1755	1357	1696	1359	-1	1742	
1561	10	2	1757	1359	1358	1743	1755	1365	1743	
1562	10	2	1359	1757	1754	1360	1756	1366	1744	
1563	10	2	1745	1362	1361	1758	-1	1369	1745	
1564	10	2	1759	1364	1363	1760	-1	1753	1384	
1565	10	2	1364	1759	1366	-1	-1	1754	1368	
1566	10	2	1365	1366	1759	1748	-1	1757	1367	
1567	10	2	1384	1368	1367	1761	-1	1747	1759	
1568	10	2	1369	1370	1371	1762	1385	1745	1383	
1569	10	2	1762	-1	1760	1369	1763	1758	1392	
1570	10	2	1372	1373	1764	1371	1765	1738	1380	
1571	10	2	1765	1374	1730	1388	1372	1766	1382	
1572	10	2	1749	1764	1373	-1	1767	1739	1381	
1573	10	2	1374	1765	1767	1390	1373	1768	1379	1376(4)
1574	10	2	1768	1766	1731	1391	1740	1374	-1	1377(4)
1575	10	1	1716	-1	1376	1769	1375	1770	1378	
1576	10	1	1377	-1	1770	1768	1736	1376	-1	
1577	10	1	1715	1771	1772	1378	-1	-1	1769	
1578	10	2	1382	1379	1773	1395	1380	-1	1765	
1579	10	2	1727	1773	1379	1772	1381	-1	1767	
1580	10	2	1747	1381	1380	1774	1773	1384	1764	
1581	10	2	1392	-1	1774	1383	1775	1392	1762	
1582	10	2	1761	-1	1392	1384	1776	1774	1760	
1583	10	2	1386	1397	1777	1385	1778	-1	1402	
1584	10	2	1778	1398	-1	1763	1386	1779	1403	
1585	10	2	-1	1399	1387	1777	1780	-1	1394	1389(2)
1586	10	2	1388	1400	1780	1765	1387	1781	1395	1390(2)
1587	10	2	1781	1401	-1	1766	-1	1388	-1	1391(2)
1588	10	1	-1	1780	1390	1782	1389	1783	1396	
1589	10	1	1391	1781	1783	1768	-1	1390	-1	
1590	10	2	1775	-1	1776	1403	1392	1784	1763	
1591	10	2	-1	1416	1404	-1	1393	1785	-1	
1592	10	2	-1	1408	1402	1394	1786	1393	1777	
1593	10	2	-1	1411	1395	1787	1394	-1	1780	1396(2)
1594	10	1	-1	1788	1772	1396	-1	-1	1782	
1595	10	2	-1	1777	1397	1405	1789	-1	1408	1399(4)
1596	10	2	1398	1778	1789	-1	1397	1790	1409	
1597	10	2	1790	1779	-1	-1	-1	1398	1791	
1598	10	1	-1	1780	1400	1792	1399	1793	1411	
1599	10	1	1401	1781	1793	-1	-1	1400	-1	
1600	10	2	1403	1409	1786	1775	1402	1794	1778	
1601	10	2	1794	1410	-1	1784	1412	1403	1795	
1602	10	2	1412	1417	1785	1412	1794	1404	1796	
1603	10	1	-1	-1	1406	1797	1405	1798	1413	
1604	10	1	1407	-1	1798	-1	-1	1406	1799	
1605	10	2	1799	-1	-1	-1	-1	1415	1425	1407(7)
1606	10	2	-1	1786	1409	1800	1408	1801	1789	
1607	10	2	1410	1794	1801	-1	1417	1409	1802	
1608	10	2	1802	1795	-1	-1	1418	1791	1410	
1609	10	1	-1	1803	-1	1411	1800	-1	1792	
1610	10	2	1796	1418	-1	1796	1795	-1	1412	
1611	10	1	-1	1804	-1	1413	1800	1805	1797	
1612	10	2	-1	-1	1414	1805	1420	1422	1806	1413(6)
1613	10	2	1415	-1	1806	-1	1421	1799	1414	
1614	10	2	-1	1785	1417	1420	1801	1416	1807	
1615	10	2	1418	1796	1807	1419	1802	-1	1417	
1616	10	2	-1	-1	1421	1807	1806	-1	1420	
1617	10	1	-1	1808	-1	1422	-1	1805	1809	
1618	10	1	-1	-1	1424	1809	1423	1810	1422	
1619	10	1	1425	-1	1810	-1	-1	1424	1799	
1620	11	1	-1	1811	-1	1426	-1	1812	1429	
1621	11	1	-1	-1	1427	1812	-1	1426	1813	
1622	11	2	1428	-1	1813	-1	-1	1431	1449	1427(7)
1623	11	1	-1	1429	-1	-1	-1	1814	1811	
1624	11	2	-1	1814	-1	1430	1815	1447	1816	1429(6)
1625	11	2	-1	-1	1431	1816	1434	1813	1430	
1626	11	2	-1	1433	1435	-1	1817	1432	1818	
1627	11	2	-1	1817	1436	1815	1433	1444	1819	
1628	11	2	-1	1818	1437	1434	1819	-1	1433	
1629	11	2	1438	1436	1817	1820	1435	1442	1821	
1630	11	2	1439	1437	1818	1439	1821	-1	1435	
1631	11	2	1440	1821	1819	-1	1437	1822	1436	
1632	11	2	1820	-1	1823	1438	1820	1453	1824	
1633	11	2	1821	1440	-1	1824	1439	1825	1438	
1634	11	2	1822	1825	-1	-1	-1	1440	1455	
1635	11	2	-1	1444	1442	1826	1441	1817	1827	
1636	11	2	-1	1445	1443	1460	1827	-1	1441	
1637	11	2	1453	-1	1828	1442	1450	1820	1829	
1638	11	2	1454	1446	1827	1829	1443	1830	1442	
1639	11	2	-1	1831	-1	1444	1463	1815	1832	1447(5)
1640	11	2	-1	1827	1446	1832	1445	1833	1444	
1641	11	2	1455	1830	1833	-1	-1	1446	1822	
1642	11	1	-1	1447	-1	-1	1831	1814	1834	
1643	11	1	-1	1834	-1	1448	1832	1835	1447	
1644	11	1	-1	-1	1449	1835	-1	1448	1813	
1645	11	2	1836	-1	1450	1469	1828	1456	1837	
1646	11	2	1457	-1	1837	1451	1829	1838	1450	
1647	11	2	1458	1452	1471	1838	-1	1451	1458	
1648	11	2	1839	-1	1453	-1	1456	1823	1840	
1649	11	2	1829	-1	1840	1454	1457	1841	1453	
1650	11	2	1830	1455	-1	1841	-1	1454	1825	
1651	11	2	1456	-1	1836	1475	1839	1836	1842	
1652	11	2	1842	-1	1457	1476	1840	1843	1456	
1653	11	2	1838	-1	1843	1458	1844	1457	1838	
1654	11	2	-1	1845	1846	1459	1826	-1	1847	
1655	11	2	-1	1482	1462	1847	1460	1848	1459	1464(2)
1656	11	2	1849	1483	1461	1846	1469	-1	1850	
1657	11	2	1467	1484	1850	1462	1470	1851	1461	
1658	11	2	1468	1485	1848	1851	-1	1462	-1	1465(2)
1659	11	1	-1	1463	-1	1845	1831	-1	1852	
1660	11	1	-1	1852	-1	1464	1832	1853	1463	
1661	11	1	-1	1848	1465	1853	-1	1464	-1	
1662	11	2	1466	1492	1849	-1	1472	-1	1854	
1663	11	2	1854	1493	1467	-1	1473	1855	1466	
1664	11	2	1851	1494	1855	1468	1856	1467	-1	

Parameter	ℓ	d	Simples 1	2	3	4	5	6	7	Other
1665	11	2	1469	1486	1472	1836	1849	1475	1857	
1666	11	2	1857	1487	1470	1837	1850	1858	1469	
1667	11	2	1474	1488	1858	1471	1859	1470	1477	
1668	11	2	1473	1495	1857	1478	1854	1860	1472	
1669	11	2	1860	1496	1474	1479	1861	1473	1479	
1670	11	2	1859	1497	1861	-1	1474	1856	1480	
1671	11	2	1476	1489	1478	1842	-1	1862	1475	
1672	11	2	1862	1490	1477	1843	1863	1476	1858	
1673	11	2	1480	1491	1863	-1	1477	-1	1859	
1674	11	2	1479	1500	1862	1860	1864	1478	1860	
1675	11	2	1864	1501	1480	1865	1479	-1	1861	
1676	11	1	-1	1481	1866	1845	-1	-1	1867	
1677	11	2	1868	1866	1505	1483	-1	-1	1869	1481(3)
1678	11	1	-1	1867	1869	1482	-1	1870	1481	
1679	11	1	-1	1848	1485	1870	-1	1482	-1	
1680	11	2	1483	1849	1492	1868	1486	-1	1871	
1681	11	2	1871	1850	1484	1869	1487	1872	1483	
1682	11	2	1494	1851	1872	1509	1873	1484	-1	1485(4)
1683	11	2	1487	1857	1495	-1	1871	1874	1486	
1684	11	2	1874	1858	1488	-1	1875	1487	1490	
1685	11	2	1497	1859	1875	1510	1488	1873	1491	
1686	11	2	1490	1862	1500	-1	1876	1489	1874	
1687	11	2	1876	1863	1491	1877	1490	-1	1875	
1688	11	2	1493	1854	1871	1513	1495	1878	1492	
1689	11	2	1878	1855	1494	1516	1879	1493	-1	
1690	11	2	1873	1856	1879	1519	1494	1497	-1	
1691	11	2	1496	1860	1874	1498	1880	1495	1500	
1692	11	2	1880	1861	1497	1881	1496	1879	1501	
1693	11	3	1499	1511	-1	1882	1546	1525	1511	1498(5)
1694	11	3	1882	1883	1526	1499	1881	1526	1512	
1695	11	2	1501	1864	1876	1884	1500	-1	1880	
1696	11	3	1884	1885	1518	1560	1512	-1	1881	1501(4)
1697	11	2	1536	1886	1868	1502	-1	-1	1887	1505(1)
1698	11	2	1537	-1	1513	1887	1503	1888	1502	1506(1)
1699	11	3	1538	-1	1514	-1	1889	1521	1504	1507(1) 1503(6)
1700	11	3	1555	-1	1515	1890	1504	-1	1889	1508(1)
1701	11	1	1886	1505	1866	-1	-1	-1	1891	
1702	11	1	1887	1891	1869	1506	-1	1892	1505	
1703	11	1	1888	-1	1509	1892	1893	1506	-1	
1704	11	2	1889	-1	1510	1894	1507	1893	1508	
1705	11	2	1890	1895	1877	1508	-1	-1	1894	
1706	11	2	1519	-1	1893	1873	1524	1510	-1	1509(5)
1707	11	3	1512	1896	1897	1511	1884	1527	1882	
1708	11	3	1896	1512	1530	1883	1885	1528	1883	
1709	11	2	1516	-1	1888	1878	1898	1513	-1	
1710	11	3	1517	-1	1889	1899	1514	1898	1515	
1711	11	3	1518	1900	1901	1515	1529	-1	1899	
1712	11	3	1898	-1	1542	1902	1535	1517	-1	1519(3) 1516(5)
1713	11	3	1899	1903	1881	1517	1526	1902	1518	
1714	11	3	1900	1518	1885	-1	1530	-1	1903	
1715	11	2	1577	1904	1905	1520	-1	-1	1906	1522(1)
1716	11	2	1575	-1	1532	1906	1521	1907	1520	1523(1)
1717	11	1	1904	1522	-1	-1	-1	-1	1908	
1718	11	1	1906	1908	-1	1523	-1	1909	1522	
1719	11	1	1907	-1	1524	1909	1893	1523	-1	
1720	11	3	1526	1910	1911	1525	1899	1882	1529	
1721	11	3	1910	1526	1883	1528	1903	1883	1530	
1722	11	3	1528	1527	1912	1910	-1	1896	1549	
1723	11	3	1550	1912	1527	-1	-1	1897	1550	
1724	11	3	1530	1529	1913	1549	1900	1549	1910	
1725	11	3	1897	1913	1529	-1	1901	1550	1911	
1726	11	2	1533	1531	1914	-1	1551	-1	1915	
1727	11	2	1579	1914	1531	1905	1552	-1	1916	
1728	11	2	1534	1915	1916	1532	1553	1917	1531	
1729	11	2	1535	-1	1907	1917	1898	1532	-1	
1730	11	2	1915	1534	1571	-1	1554	1918	1533	
1731	11	2	1917	1918	1574	1535	1919	1534	-1	
1732	11	1	1886	1536	-1	-1	-1	-1	1920	
1733	11	1	1887	1920	-1	1537	-1	1921	1536	
1734	11	1	1888	-1	1539	1921	1922	1537	-1	
1735	11	2	1889	-1	1540	1923	1538	1922	1555	
1736	11	2	1542	-1	1922	1924	1576	1540	-1	1539(5)
1737	11	2	1543	1925	1926	1540	1541	1924	1556	
1738	11	2	1544	1541	1927	1547	1925	1570	1557	
1739	11	2	1928	1927	1541	-1	1926	1572	1558	
1740	11	2	1924	1929	1919	1542	1574	1543	-1	
1741	11	2	1925	1543	1930	-1	1544	1929	1559	
1742	11	2	1931	1930	1543	1881	1545	1919	1560	
1743	11	2	1932	1545	1544	1561	1930	1554	1561	
1744	11	2	1545	1932	1928	1546	1931	1553	1562	
1745	11	2	1563	1548	1547	1933	-1	1568	1563	
1746	11	3	1912	1550	1549	1934	-1	1913	1912	
1747	11	2	1580	1552	1551	1935	1914	1567	1936	
1748	11	2	1554	1553	1936	1566	1915	1932	1551	
1749	11	2	1572	1936	1553	-1	1916	1928	1552	
1750	11	2	1890	1937	1938	1555	-1	-1	1923	
1751	11	2	1559	1556	1939	-1	1557	-1	1925	
1752	11	2	1940	1939	1556	1938	1558	-1	1926	
1753	11	2	1941	1558	1557	1942	1939	1564	1927	
1754	11	2	1558	1941	1562	-1	1940	1565	1928	
1755	11	2	1943	1560	1559	1944	1561	-1	1930	
1756	11	2	1560	1943	1940	1884	1562	-1	1931	
1757	11	2	1561	1562	1941	1932	1943	1566	1932	
1758	11	2	1933	-1	1942	1563	1945	1569	1933	
1759	11	2	1564	1565	1566	1946	-1	1941	1567	
1760	11	2	1946	-1	1569	1564	1947	1942	1582	
1761	11	2	1582	-1	1935	1567	1948	1935	1946	
1762	11	2	1569	-1	1949	1568	1950	1933	1581	
1763	11	2	1950	-1	1947	1584	1569	1951	1590	
1764	11	2	1936	1572	1570	1949	1952	1927	1580	
1765	11	2	1571	1573	1952	1586	1570	1953	1578	
1766	11	2	1953	1574	1918	1587	1929	1571	-1	
1767	11	2	1916	1952	1573	1954	1572	1955	1579	
1768	11	2	1574	1953	1955	1589	1924	1573	-1	1576(4)
1769	11	1	1906	1956	1954	1575	-1	1957	1577	
1770	11	1	1907	-1	1576	1957	1922	1575	-1	
1771	11	1	1904	1577	1958	-1	-1	-1	1956	
1772	11	2	1905	1958	1594	1579	-1	-1	1954	1577(3)
1773	11	2	1914	1579	1578	1959	1580	-1	1952	
1774	11	2	1935	-1	1581	1580	1960	1582	1949	
1775	11	2	1590	-1	1960	1600	1581	1961	1950	
1776	11	2	1948	-1	1590	-1	1582	1962	1947	
1777	11	2	-1	1595	1583	1585	1963	-1	1592	
1778	11	2	1584	1596	1963	1950	1583	1964	1600	
1779	11	2	1964	1597	-1	1951	-1	1584	1965	
1780	11	2	-1	1598	1586	1966	1585	1967	1593	1588(2)
1781	11	2	1587	1599	1967	1953	-1	1586	-1	1589(2)
1782	11	1	-1	1968	1954	1588	-1	1969	1594	
1783	11	1	-1	1967	1589	1969	-1	1588	-1	
1784	11	2	1961	-1	1962	1601	1961	1590	1970	
1785	11	2	-1	1614	1602	-1	1971	1591	1972	
1786	11	2	-1	1606	1600	1973	1592	1971	1963	
1787	11	2	-1	1974	1959	1593	1973	-1	1966	
1788	11	1	-1	1594	1958	1974	-1	-1	1968	
1789	11	2	-1	1963	1596	1975	1595	1976	1606	
1790	11	2	1597	1964	1976	-1	-1	1596	1977	
1791	11	2	1977	1965	-1	-1	-1	1608	1597	
1792	11	1	-1	1978	-1	1598	1975	1979	1609	
1793	11	1	-1	1967	1599	1979	-1	1598	-1	
1794	11	2	1601	1607	1971	1961	1602	1600	1980	
1795	11	2	1980	1608	-1	1970	1610	1965	1601	
1796	11	2	1610	1615	1972	1610	1980	-1	1602	
1797	11	1	-1	1981	-1	1603	1975	1982	1611	
1798	11	1	-1	-1	1604	1982	-1	1603	1983	
1799	11	2	1605	-1	1983	-1	-1	1613	1619	1604(7)
1800	11	2	-1	1984	-1	1606	1611	1985	1975	1609(5)
1801	11	2	-1	1971	1607	1985	1614	1606	1986	
1802	11	2	1608	1980	1986	-1	1615	1977	1607	
1803	11	1	-1	1609	-1	1974	1984	-1	1978	
1804	11	1	-1	1611	-1	-1	1984	1987	1981	
1805	11	2	-1	1987	-1	1612	1985	1617	1988	1611(6)
1806	11	2	-1	-1	1613	1988	1616	1983	1612	
1807	11	2	-1	1972	1615	1616	1986	-1	1614	
1808	11	1	-1	1617	-1	-1	-1	1987	1989	
1809	11	1	-1	1989	-1	1618	-1	1990	1617	
1810	11	1	-1	-1	1619	1990	-1	1618	1983	
1811	12	1	-1	1620	-1	-1	-1	1991	1623	
1812	12	1	-1	1991	-1	1621	1992	1620	1993	
1813	12	2	-1	-1	1622	1993	-1	1625	1644	1621(7)
1814	12	2	-1	1624	-1	-1	1994	1642	1995	1623(6)
1815	12	2	-1	1994	-1	1627	1624	1639	1996	
1816	12	2	-1	1995	-1	1625	1996	1993	1624	

Para-meter	ℓ	d	Simples 1	2	3	4	5	6	7	Other
1817	12	2	-1	1627	1629	1997	1626	1635	1998	
1818	12	2	-1	1628	1630	-1	1998	-1	1626	
1819	12	2	-1	1998	1631	1996	1628	1999	1627	
1820	12	2	1632	-1	2000	1629	1632	1637	2001	
1821	12	2	1633	1631	1998	2001	1630	2002	1629	
1822	12	2	1634	2002	1999	-1	-1	1631	1641	
1823	12	2	2003	-1	1632	-1	2000	1648	2004	
1824	12	2	2001	-1	2004	1633	2001	2005	1632	
1825	12	2	2002	1634	-1	2005	-1	1633	1650	
1826	12	2	-1	2006	2007	1635	1654	1997	2008	
1827	12	2	-1	1640	1638	2008	1636	2009	1635	
1828	12	2	2010	-1	1637	2007	1645	2000	2011	
1829	12	2	1649	-1	2011	1638	1646	2012	1637	
1830	12	2	1650	1641	2009	2012	-1	1638	2002	
1831	12	2	-1	1639	-1	2006	1659	1994	2013	1642(5)
1832	12	2	-1	2013	-1	1640	1660	2014	1639	1643(5)
1833	12	2	-1	2009	1641	2014	-1	1640	1999	
1834	12	1	-1	1643	-1	-1	2013	2015	1642	
1835	12	1	-1	2015	-1	1644	2016	1643	1993	
1836	12	2	1645	-1	1651	1665	2010	1651	2017	
1837	12	2	2017	-1	1646	1666	2011	2018	1645	
1838	12	2	1653	-1	2018	1647	2019	1646	1653	
1839	12	2	1648	-1	2010	-1	1651	2003	2020	
1840	12	2	2020	-1	1649	-1	1652	2021	1648	
1841	12	2	2012	-1	2021	1650	2022	1649	2005	
1842	12	2	1652	-1	2017	1671	2020	2023	1651	
1843	12	2	2023	-1	1653	1672	2024	1652	2018	
1844	12	2	2019	-1	2024	-1	1653	2022	2019	
1845	12	2	-1	1654	2025	1676	2006	-1	2026	1659(4)
1846	12	2	2027	2025	1654	1656	2007	-1	2028	
1847	12	2	-1	2026	2028	1655	2008	2029	1654	
1848	12	2	-1	1679	1658	2029	-1	1655	-1	1661(2)
1849	12	2	1656	1680	1662	2027	1665	-1	2030	
1850	12	2	2030	1681	1657	2028	1666	2031	1656	
1851	12	2	1664	1682	2031	1658	2032	1657	-1	
1852	12	1	-1	1660	-1	2026	2013	2033	1659	
1853	12	1	-1	2033	-1	1661	2034	1660	-1	
1854	12	2	1663	1688	2030	-1	1668	2035	1662	
1855	12	2	2035	1689	1664	-1	2036	1663	-1	
1856	12	2	2032	1690	2036	-1	1664	1670	-1	
1857	12	2	1666	1683	1668	2017	2030	2037	1665	
1858	12	2	2037	1684	1667	2018	2038	1666	1672	
1859	12	2	1670	1685	2038	-1	1667	2032	1673	
1860	12	2	1669	1691	2037	1674	2039	1668	1674	
1861	12	2	2039	1692	1670	2040	1669	2036	1675	
1862	12	2	1672	1686	1674	2023	2041	1671	2037	
1863	12	2	2041	1687	1673	2042	1672	-1	2038	
1864	12	2	1675	1695	2041	2043	1674	-1	2039	
1865	12	2	2043	2044	-1	1675	2040	-1	2040	
1866	12	2	2045	1677	1701	2025	-1	-1	2046	1676(3)
1867	12	1	-1	1678	2046	2026	-1	2047	1676	
1868	12	2	1677	2045	1697	1680	-1	-1	2048	
1869	12	2	2048	2046	1702	1681	-1	2049	1677	1678(3)
1870	12	1	-1	2047	2049	1679	2050	1678	-1	
1871	12	2	1681	2030	1688	2048	1683	2051	1680	
1872	12	2	2051	2031	1682	2049	2052	1681	-1	
1873	12	2	1690	2032	2052	1706	1682	1685	-1	
1874	12	2	1684	2037	1691	-1	2053	1683	1686	
1875	12	2	2053	2038	1685	2054	1684	2052	1687	
1876	12	2	1687	2041	1695	2055	1686	-1	2053	
1877	12	2	2055	2056	1705	1687	-1	-1	2054	
1878	12	2	1689	2035	2051	1709	2057	1688	-1	
1879	12	2	2057	2036	1690	2058	1689	1692	-1	
1880	12	2	1692	2039	2053	2059	1691	2057	1695	
1881	12	3	2059	2060	1713	1742	1694	2058	1696	1692(4)
1882	12	3	1694	2061	2062	1693	2059	1720	1707	
1883	12	3	2061	1694	1721	1708	2060	1721	1708	
1884	12	3	1696	2063	2064	1756	1707	-1	2059	1695(4)
1885	12	3	2063	1696	1714	2044	1708	-1	2060	
1886	12	2	1732	1697	2045	-1	-1	-1	2065	1701(1)
1887	12	2	1733	2065	2048	1698	-1	2066	1697	1702(1)
1888	12	2	1734	-1	1709	2066	2067	1698	-1	1703(1)
1889	12	3	1735	-1	1710	2068	1699	2067	1700	1704(1)
1890	12	3	1750	2069	2070	1700	-1	-1	2068	1705(1)
1891	12	1	2065	1702	2046	-1	-1	2071	1701	
1892	12	1	2066	2071	2049	1703	2072	1702	-1	
1893	12	2	2067	-1	1706	2073	1719	1704	-1	1703(5)
1894	12	2	2068	2074	2054	1704	-1	2073	1705	
1895	12	2	2069	1705	2056	-1	-1	-1	2074	
1896	12	3	1708	1707	2075	2061	2063	1722	2061	
1897	12	3	1725	2075	1707	-1	2064	1723	2062	
1898	12	3	1712	-1	2067	2076	1729	1710	-1	1709(5)
1899	12	3	1713	2077	2078	1710	1720	2076	1711	
1900	12	3	1714	1711	2079	-1	1724	-1	2077	
1901	12	3	2064	2079	1711	2070	1725	-1	2078	
1902	12	3	2076	2080	2058	1712	2058	1713	-1	
1903	12	3	2077	1713	2060	-1	1721	2080	1714	
1904	12	2	1771	1715	2081	-1	-1	-1	2082	1717(1)
1905	12	2	1772	2081	1715	1727	-1	-1	2083	
1906	12	2	1769	2082	2083	1716	-1	2084	1715	1718(1)
1907	12	2	1770	-1	1729	2084	2067	1716	-1	1719(1)
1908	12	1	2082	1718	-1	-1	-1	2085	1717	
1909	12	1	2084	2085	-1	1719	2086	1718	-1	
1910	12	3	1721	1720	2087	1722	2077	2061	1724	
1911	12	3	2062	2087	1720	-1	2078	2062	1725	
1912	12	3	1746	1723	1722	2088	-1	2075	1746	
1913	12	3	2075	1725	1724	2089	2079	1746	2087	
1914	12	2	1773	1727	1726	2090	1747	-1	2091	
1915	12	2	1730	1728	2091	-1	1748	2092	1726	
1916	12	2	1767	2091	1728	2083	1749	2093	1727	
1917	12	2	1731	2092	2093	1729	2094	1728	-1	
1918	12	2	2092	1731	1766	-1	2095	1730	-1	
1919	12	2	2094	2095	1740	2058	1731	1742	-1	
1920	12	1	2065	1733	-1	-1	-1	2096	1732	
1921	12	1	2066	2096	-1	1734	2097	1733	-1	
1922	12	2	2067	-1	1736	2098	1770	1735	-1	1734(5)
1923	12	2	2068	2099	2100	1735	-1	2098	1750	
1924	12	2	1740	2101	2102	1736	1768	1737	-1	
1925	12	2	1741	1737	2103	-1	1738	2101	1751	
1926	12	2	2104	2103	1737	2100	1739	2102	1752	
1927	12	2	2105	1739	1738	2106	2103	1764	1753	
1928	12	2	1739	2105	1744	-1	2104	1749	1754	
1929	12	2	2101	1740	2095	-1	1766	1741	-1	
1930	12	2	2107	1742	1741	2108	1743	2095	1755	
1931	12	2	1742	2107	2104	2059	1744	2094	1756	
1932	12	2	1743	1744	2105	1757	2107	1748	1757	
1933	12	2	1758	-1	2106	1745	2109	1762	1758	
1934	12	3	2088	-1	2089	1746	2110	2089	2088	
1935	12	2	1774	-1	1761	1747	2111	1761	2112	
1936	12	2	1764	1749	1748	2112	2091	2105	1747	
1937	12	2	2069	1750	2113	-1	-1	-1	2099	
1938	12	2	2114	2113	1750	1752	-1	-1	2100	
1939	12	2	2115	1752	1751	2116	1753	-1	2103	
1940	12	2	1752	2115	1756	2114	1754	-1	2104	
1941	12	2	1753	1754	1757	2117	2115	1759	2105	
1942	12	2	2117	-1	1758	1753	2118	1760	2106	
1943	12	2	1755	1756	2115	2119	1757	-1	2107	
1944	12	2	2119	2044	-1	1755	2108	-1	2108	
1945	12	2	2109	-1	2118	-1	1758	2120	2109	
1946	12	2	1760	-1	2112	1759	2121	2117	1761	
1947	12	2	2121	-1	1763	-1	1760	2122	1776	
1948	12	2	1776	-1	2111	-1	1761	2123	2121	
1949	12	2	2112	-1	1762	1764	2124	2106	1774	
1950	12	2	1763	-1	2124	1778	1762	2125	1775	
1951	12	2	2125	-1	2122	1779	2120	1763	2126	
1952	12	2	2091	1767	1765	2127	1764	2128	1773	
1953	12	2	1766	1768	2128	1781	2101	1765	-1	
1954	12	2	2083	2129	1782	1767	-1	2130	1772	1769(3)
1955	12	2	2093	2128	1768	2130	2102	1767	-1	
1956	12	1	2082	1769	2129	-1	-1	2131	1771	
1957	12	1	2084	2131	2130	1770	2132	1769	-1	
1958	12	2	2081	1772	1788	2133	-1	-1	2129	1771(3)
1959	12	2	2090	2133	1787	1773	2134	-1	2127	
1960	12	2	2111	-1	1775	2134	1774	2135	2124	
1961	12	2	1784	-1	2135	1794	1784	1775	2136	
1962	12	2	2123	-1	1784	-1	2135	1776	2137	
1963	12	2	-1	1789	1778	2138	1777	2139	1786	
1964	12	2	1779	1790	2139	2125	-1	1778	2140	
1965	12	2	2140	1791	-1	2126	-1	1795	1779	
1966	12	2	-1	2141	2127	1780	2138	2142	1787	
1967	12	2	-1	1793	1781	2142	-1	1780	-1	1783(2)
1968	12	1	-1	1782	2129	2141	-1	2143	1788	

Parameter	ℓ	d	Simples 1	2	3	4	5	6	7	Other
1969	12	1	-1	2143	2130	1783	2144	1782	-1	
1970	12	2	2136	-1	2137	1795	2136	2126	1784	
1971	12	2	-1	1801	1794	2145	1785	1786	2146	
1972	12	2	-1	1807	1796	-1	2146	-1	1785	
1973	12	2	-1	2147	2134	1786	1787	2145	2138	
1974	12	2	-1	1787	2133	1803	2147	-1	2141	1788(4)
1975	12	2	-1	2148	-1	1789	1797	2149	1800	1792(5)
1976	12	2	-1	2139	1790	2149	-1	1789	2150	
1977	12	2	1791	2140	2150	-1	-1	1802	1790	
1978	12	1	-1	1792	-1	2141	2148	2151	1803	
1979	12	1	-1	2151	-1	1793	2152	1792	-1	
1980	12	2	1795	1802	2146	2136	1796	2140	1794	
1981	12	1	-1	1797	-1	-1	2148	2153	1804	
1982	12	1	-1	2153	-1	1798	2154	1797	2155	
1983	12	2	-1	-1	1799	2155	-1	1806	1810	1798(7)
1984	12	2	-1	1800	-1	2147	1804	2156	2148	1803(5)
1985	12	2	-1	2156	-1	1801	1805	1800	2157	
1986	12	2	-1	2146	1802	2157	1807	2150	1801	
1987	12	2	-1	1805	-1	-1	2156	1808	2158	1804(6)
1988	12	2	-1	2158	-1	1806	2157	2155	1805	
1989	12	1	-1	1809	-1	-1	-1	2159	1808	
1990	12	1	-1	2159	-1	1810	2160	1809	2155	
1991	13	1	-1	1812	-1	-1	2161	1811	2162	
1992	13	1	-1	2161	-1	-1	1812	-1	2163	
1993	13	2	-1	2162	-1	1813	2163	1816	1835	1812(7)
1994	13	2	-1	1815	-1	2164	1814	1831	2165	
1995	13	2	-1	1816	-1	-1	2165	2162	1814	
1996	13	2	-1	2165	-1	1819	1816	2166	1815	
1997	13	2	-1	2164	2167	1817	-1	1826	2168	
1998	13	2	-1	1819	1821	2168	1818	2169	1817	
1999	13	2	-1	2169	1822	2166	-1	1819	1833	
2000	13	2	2170	-1	1820	2167	1823	1828	2171	
2001	13	2	1824	-1	2171	1821	1824	2172	1820	
2002	13	2	1825	1822	2169	2172	-1	1821	1830	
2003	13	2	1823	-1	2170	-1	2170	1839	2173	
2004	13	2	2173	-1	1824	-1	2171	2174	1823	
2005	13	2	2172	-1	2174	1825	2175	1824	1841	
2006	13	2	-1	1826	2176	1831	1845	2164	2177	
2007	13	2	2178	2176	1826	1828	1846	2167	2179	
2008	13	2	-1	2177	2179	1827	1847	2180	1826	
2009	13	2	-1	1833	1830	2180	-1	1827	2169	
2010	13	2	1828	-1	1839	2178	1836	2170	2181	
2011	13	2	2181	-1	1829	2179	1837	2182	1828	
2012	13	2	1841	-1	2182	1830	2183	1829	2172	
2013	13	2	-1	1832	-1	2177	1852	2184	1831	1834(5)
2014	13	2	-1	2184	-1	1833	2185	1832	2166	
2015	13	1	-1	1835	-1	-1	2186	1834	2162	
2016	13	1	-1	2186	-1	-1	1835	2185	2163	
2017	13	2	1837	-1	1842	1857	2181	2187	1836	
2018	13	2	2187	-1	1838	1858	2188	1837	1843	
2019	13	2	1844	-1	2188	-1	1838	2183	1844	
2020	13	2	1840	-1	2181	-1	1842	2189	1839	
2021	13	2	2189	-1	1841	-1	2190	1840	2174	
2022	13	2	2183	-1	2190	-1	1841	1844	2175	
2023	13	2	1843	-1	2187	1862	2191	1842	2187	
2024	13	2	2191	-1	1844	2192	1843	2190	2188	
2025	13	2	2193	1846	1845	1866	2176	-1	2194	
2026	13	2	-1	1847	2194	1867	2177	2195	1845	1852(4)
2027	13	2	1846	2193	-1	1849	2178	-1	2196	
2028	13	2	2196	2194	1847	1850	2179	2197	1846	
2029	13	2	-1	2195	2197	1848	2198	1847	-1	
2030	13	2	1850	1871	1854	2196	1857	2199	1849	
2031	13	2	2199	1872	1851	2197	2200	1850	-1	
2032	13	2	1856	1873	2200	-1	1851	1859	-1	
2033	13	1	-1	1853	-1	2195	2201	1852	-1	
2034	13	1	-1	2201	-1	-1	1853	2185	-1	
2035	13	2	1855	1878	2199	-1	2202	1854	-1	
2036	13	2	2202	1879	1856	2203	1855	1861	-1	
2037	13	2	1858	1874	1860	2187	2204	1857	1862	
2038	13	2	2204	1875	1859	2205	1858	2200	1863	
2039	13	2	1861	1880	2204	2206	1860	2202	1864	
2040	13	2	2206	2207	-1	1861	1865	2203	1865	
2041	13	2	1863	1876	1864	2208	1862	-1	2204	
2042	13	2	2208	2209	-1	1863	2192	-1	2205	
2043	13	2	1865	2210	2211	1864	2206	-1	2206	
2044	13	3	2210	1944	-1	1885	2207	-1	2207	1865(2)
2045	13	2	1866	1868	1886	2193	-1	-1	2212	
2046	13	2	2212	1869	1891	2194	-1	2213	1866	1867(3)
2047	13	1	-1	1870	2213	2195	2214	1867	-1	
2048	13	2	1869	2212	1887	1871	-1	2215	1868	
2049	13	2	2215	2213	1892	1872	2216	1869	-1	1870(3)
2050	13	1	-1	2214	2216	-1	1870	-1	-1	
2051	13	2	1872	2199	1878	2215	2217	1871	-1	
2052	13	2	2217	2200	1873	2218	1872	1875	-1	
2053	13	2	1875	2204	1880	2219	1874	2217	1876	
2054	13	2	2219	2220	1894	1875	-1	2218	1877	
2055	13	2	1877	2221	2222	1876	-1	-1	2219	
2056	13	2	2221	1877	1895	2209	-1	-1	2220	
2057	13	2	1879	2202	2217	2223	1878	1880	-1	
2058	13	3	2223	2224	1902	1919	1902	1881	-1	1879(4)
2059	13	3	1881	2225	2226	1931	1882	2223	1884	1880(4)
2060	13	3	2225	1881	1903	2207	1883	2224	1885	
2061	13	3	1883	1882	2227	1896	2225	1910	1896	
2062	13	3	1911	2227	1882	-1	2226	1911	1897	
2063	13	3	1885	1884	2228	2210	1896	-1	2225	
2064	13	3	1901	2228	1884	2222	1897	-1	2226	
2065	13	2	1920	1887	2212	-1	-1	2229	1886	1891(1)
2066	13	2	1921	2229	2215	1888	2230	1887	-1	1892(1)
2067	13	3	1922	-1	1898	2231	1907	1889	-1	1893(1) 1888(5)
2068	13	3	1923	2232	2233	1889	-1	2231	1890	1894(1)
2069	13	3	1937	1890	2234	-1	-1	-1	2232	1895(1)
2070	13	3	2222	2234	1890	1901	-1	-1	2233	
2071	13	1	2229	1892	2213	-1	2235	1891	-1	
2072	13	1	2230	2235	2216	2236	1892	-1	-1	
2073	13	2	2231	2237	2218	1893	2236	1894	-1	
2074	13	2	2232	1894	2220	-1	-1	2237	1895	
2075	13	3	1913	1897	1896	2238	2228	1912	2227	
2076	13	3	1902	2239	2240	1898	2223	1899	-1	
2077	13	3	1903	1899	2241	-1	1910	2239	1900	
2078	13	3	2226	2241	1899	2233	1911	2240	1901	
2079	13	3	2228	1901	1900	2242	1913	-1	2241	
2080	13	3	2239	1902	2224	-1	2224	1903	-1	
2081	13	2	1958	1905	1904	2243	-1	-1	2244	
2082	13	2	1956	1906	2244	-1	-1	2245	1904	1908(1)
2083	13	2	1954	2244	1906	1916	-1	2246	1905	
2084	13	2	1957	2245	2246	1907	2247	1906	-1	1909(1)
2085	13	1	2245	1909	-1	-1	2248	1908	-1	
2086	13	1	2247	2248	-1	2236	1909	-1	-1	
2087	13	3	2227	1911	1910	2249	2241	2227	1913	
2088	13	3	1934	-1	2249	1912	2250	2238	1934	
2089	13	3	2238	-1	1934	1913	2251	1934	2249	
2090	13	2	1959	2243	-1	1914	2252	-1	2253	
2091	13	2	1952	1916	1915	2253	1936	2254	1914	
2092	13	2	1918	1917	2254	-1	2255	1915	-1	
2093	13	2	1955	2254	1917	2246	2256	1916	-1	
2094	13	2	1919	2255	2256	2223	1917	1931	-1	
2095	13	2	2255	1919	1929	2257	1918	1930	-1	
2096	13	1	2229	1921	-1	-1	2258	1920	-1	
2097	13	1	2230	2258	-1	2259	1921	-1	-1	
2098	13	2	2231	2260	2261	1922	2259	1923	-1	
2099	13	2	2232	1923	2262	-1	-1	2260	1937	
2100	13	2	2263	2262	1923	1926	-1	2261	1938	
2101	13	2	1929	1924	2264	-1	1953	1925	-1	
2102	13	2	2256	2264	1924	2261	1955	1926	-1	
2103	13	2	2265	1926	1925	2266	1927	2264	1939	
2104	13	2	1926	2265	1931	2263	1928	2256	1940	
2105	13	2	1927	1928	1932	2267	2265	1936	1941	
2106	13	2	2267	-1	1933	1927	2268	1949	1942	
2107	13	2	1930	1931	2265	2269	1932	2255	1943	
2108	13	2	2269	2207	-1	1930	1944	2257	1944	
2109	13	2	1945	-1	2268	-1	1933	2270	1945	
2110	13	3	2250	-1	2251	-1	1934	2271	2250	
2111	13	2	1960	-1	1948	2252	1935	2272	2273	
2112	13	2	1949	-1	1946	1936	2273	2267	1935	
2113	13	2	2274	1938	1937	2275	-1	-1	2262	
2114	13	2	1938	2274	2222	1940	-1	-1	2263	
2115	13	2	1939	1940	1943	2276	1941	-1	2265	
2116	13	2	2276	2275	-1	1939	2277	-1	2266	
2117	13	2	1942	-1	2267	1941	2278	1946	2267	
2118	13	2	2278	-1	1945	2277	1942	2279	2268	
2119	13	2	1944	2210	2280	1943	2269	-1	2269	
2120	13	2	2270	-1	2279	-1	1951	1945	2281	

Parameter	ℓ	d	Simples 1	2	3	4	5	6	7	Other
2121	13	2	1947	-1	2273	-1	1946	2282	1948	
2122	13	2	2282	-1	1951	-1	2279	1947	2283	
2123	13	2	1962	-1	2272	-1	2272	1948	2284	
2124	13	2	2273	-1	1950	2285	1949	2286	1960	
2125	13	2	1951	-1	2286	1964	2270	1950	2287	
2126	13	2	2287	-1	2283	1965	2281	1970	1951	
2127	13	2	2253	2288	1966	1952	2285	2289	1959	
2128	13	2	2254	1955	1953	2289	2264	1952	-1	
2129	13	2	2244	1954	1968	2288	-1	2290	1958	1956(3)
2130	13	2	2246	2290	1969	1955	2291	1954	-1	1957(3)
2131	13	1	2245	1957	2290	-1	2292	1956	-1	
2132	13	1	2247	2292	2291	2259	1957	-1	-1	
2133	13	2	2243	1959	1974	1958	2293	-1	2288	
2134	13	2	2252	2293	1973	1960	1959	2294	2285	
2135	13	2	2272	-1	1961	2294	1962	1960	2295	
2136	13	2	1970	-1	2295	1980	1970	2287	1961	
2137	13	2	2284	-1	1970	-1	2295	2283	1962	
2138	13	2	-1	2296	2285	1963	1966	2297	1973	
2139	13	2	-1	1976	1964	2297	-1	1963	2298	
2140	13	2	1965	1977	2298	2287	-1	1980	1964	
2141	13	2	-1	1966	2288	1978	2296	2299	1974	1968(4)
2142	13	2	-1	2299	2289	1967	2300	1966	-1	
2143	13	1	-1	1969	2290	2299	2301	1968	-1	
2144	13	1	-1	2301	2291	-1	1969	-1	-1	
2145	13	2	-1	2302	2294	1971	-1	1973	2303	
2146	13	2	-1	1986	1980	2303	1972	2298	1971	
2147	13	2	-1	1973	2293	1984	1974	2302	2296	
2148	13	2	-1	1975	-1	2296	1981	2304	1984	1978(5)
2149	13	2	-1	2304	-1	1976	2305	1975	2306	
2150	13	2	-1	2298	1977	2306	-1	1986	1976	
2151	13	1	-1	1979	-1	2299	2307	1978	-1	
2152	13	1	-1	2307	-1	-1	1979	2305	-1	
2153	13	1	-1	1982	-1	-1	2308	1981	2309	
2154	13	1	-1	2308	-1	-1	1982	2305	2310	
2155	13	2	-1	2309	-1	1983	2310	1988	1990	1982(7)
2156	13	2	-1	1985	-1	2302	1987	1984	2311	
2157	13	2	-1	2311	-1	1986	1988	2306	1985	
2158	13	2	-1	1988	-1	-1	2311	2309	1987	
2159	13	1	-1	1990	-1	-1	2312	1989	2309	
2160	13	1	-1	2312	-1	-1	1990	-1	2310	
2161	14	1	-1	1992	-1	2313	1991	-1	2314	
2162	14	2	-1	1993	-1	-1	2314	1995	2015	1991(7)
2163	14	2	-1	2314	-1	-1	1993	2315	2016	1992(7)
2164	14	2	-1	1997	2316	1994	-1	2006	2317	
2165	14	2	-1	1996	-1	2317	1995	2318	1994	
2166	14	2	-1	2318	-1	1999	2315	1996	2014	
2167	14	2	2319	2316	1997	2000	-1	2007	2320	
2168	14	2	-1	2317	2320	1998	-1	2321	1997	
2169	14	2	-1	1999	2002	2321	-1	1998	2009	
2170	14	2	2000	-1	2003	2319	2003	2010	2322	
2171	14	2	2322	-1	2001	2320	2004	2323	2000	
2172	14	2	2005	-1	2323	2002	2324	2001	2012	
2173	14	2	2004	-1	2322	-1	2322	2325	2003	
2174	14	2	2325	-1	2005	-1	2326	2004	2021	
2175	14	2	2324	-1	2326	-1	2005	2324	2022	
2176	14	2	2327	2007	2006	-1	2025	2316	2328	
2177	14	2	-1	2008	2328	2013	2026	2329	2006	
2178	14	2	2007	2327	-1	2010	2027	2319	2330	
2179	14	2	2330	2328	2008	2011	2028	2331	2007	
2180	14	2	-1	2329	2331	2009	2332	2008	2321	
2181	14	2	2011	-1	2020	2330	2017	2333	2010	
2182	14	2	2333	-1	2012	2331	2334	2011	2323	
2183	14	2	2022	-1	2334	-1	2012	2019	2324	
2184	14	2	-1	2014	-1	2329	2335	2013	2318	
2185	14	2	-1	2335	-1	-1	2014	2034	2315	2016(6)
2186	14	1	-1	2016	-1	2336	2015	2335	2314	
2187	14	2	2018	-1	2023	2037	2337	2017	2023	
2188	14	2	2337	-1	2019	2338	2018	2334	2024	
2189	14	2	2021	-1	2333	-1	2339	2020	2325	
2190	14	2	2339	-1	2022	2340	2021	2024	2326	
2191	14	2	2024	-1	2337	2341	2023	2339	2337	
2192	14	2	2341	2342	-1	2024	2042	2340	2338	
2193	14	2	2025	2027	-1	2045	2327	-1	2343	
2194	14	2	2343	2028	2026	2046	2328	2344	2025	
2195	14	2	-1	2029	2344	2047	2345	2026	-1	2033(4)
2196	14	2	2028	2343	-1	2030	2330	2346	2027	
2197	14	2	2346	2344	2029	2031	2347	2028	-1	
2198	14	2	-1	2345	2347	-1	2029	2332	-1	
2199	14	2	2031	2051	2035	2346	2348	2030	-1	
2200	14	2	2348	2052	2032	2349	2031	2038	-1	
2201	14	1	-1	2034	-1	2350	2033	2335	-1	
2202	14	2	2036	2057	2348	2351	2035	2039	-1	
2203	14	2	2351	2352	-1	2036	-1	2040	-1	
2204	14	2	2038	2053	2039	2353	2037	2348	2041	
2205	14	2	2353	2354	-1	2038	2338	2349	2042	
2206	14	2	2040	2355	2356	2039	2043	2351	2043	
2207	14	3	2355	2108	-1	2060	2044	2352	2044	2040(2)
2208	14	2	2042	2357	2358	2041	2341	-1	2353	
2209	14	2	2357	2042	-1	2056	2342	-1	2354	
2210	14	3	2044	2119	2359	2063	2355	-1	2355	2043(2)
2211	14	2	-1	2359	2043	2358	2356	-1	2356	
2212	14	2	2046	2048	2065	2343	-1	2360	2045	
2213	14	2	2360	2049	2071	2344	2361	2046	-1	2047(3)
2214	14	1	-1	2050	2361	2362	2047	-1	-1	
2215	14	2	2049	2360	2066	2051	2363	2048	-1	
2216	14	2	2363	2361	2072	2364	2049	-1	-1	2050(3)
2217	14	2	2052	2348	2057	2365	2051	2053	-1	
2218	14	2	2365	2366	2073	2052	2364	2054	-1	
2219	14	2	2054	2367	2368	2053	-1	2365	2055	
2220	14	2	2367	2054	2074	2354	-1	2366	2056	
2221	14	2	2056	2055	2369	2357	-1	-1	2367	
2222	14	3	2070	2369	2114	2064	-1	-1	2368	2055(3)
2223	14	3	2058	2370	2371	2094	2076	2059	-1	2057(4)
2224	14	3	2370	2058	2080	2352	2080	2060	-1	
2225	14	3	2060	2059	2372	2355	2061	2370	2063	
2226	14	3	2078	2372	2059	2368	2062	2371	2064	
2227	14	3	2087	2062	2061	2373	2372	2087	2075	
2228	14	3	2079	2064	2063	2374	2075	-1	2372	
2229	14	2	2096	2066	2360	-1	2375	2065	-1	2071(1)
2230	14	2	2097	2375	2363	2376	2066	-1	-1	2072(1)
2231	14	3	2098	2377	2378	2067	2376	2068	-1	2073(1)
2232	14	3	2099	2068	2379	-1	-1	2377	2069	2074(1)
2233	14	3	2368	2379	2068	2078	-1	2378	2070	
2234	14	3	2369	2070	2069	2380	-1	-1	2379	
2235	14	1	2375	2072	2361	2381	2071	-1	-1	
2236	14	2	2376	2382	2364	2086	2073	-1	-1	2072(4)
2237	14	2	2377	2073	2366	-1	2382	2074	-1	
2238	14	3	2089	-1	2373	2075	2383	2088	2373	
2239	14	3	2080	2076	2384	-1	2370	2077	-1	
2240	14	3	2371	2384	2076	2378	2371	2078	-1	
2241	14	3	2372	2078	2077	2385	2087	2384	2079	
2242	14	3	2374	2380	-1	2079	2386	-1	2385	
2243	14	2	2133	2090	-1	2081	2387	-1	2388	
2244	14	2	2129	2083	2082	2388	-1	2389	2081	
2245	14	2	2131	2084	2389	-1	2390	2082	-1	2085(1)
2246	14	2	2130	2389	2084	2093	2391	2083	-1	
2247	14	2	2132	2390	2391	2376	2084	-1	-1	2086(1)
2248	14	1	2390	2086	-1	2392	2085	-1	-1	
2249	14	3	2373	-1	2088	2087	2393	2373	2089	
2250	14	3	2110	-1	2393	-1	2088	2394	2110	
2251	14	3	2383	-1	2110	2386	2089	2395	2393	
2252	14	2	2134	2387	-1	2111	2090	2396	2397	
2253	14	2	2127	2388	-1	2091	2397	2398	2090	
2254	14	2	2128	2093	2092	2398	2399	2091	-1	
2255	14	2	2095	2094	2399	2400	2092	2107	-1	
2256	14	2	2102	2399	2094	2401	2093	2104	-1	
2257	14	2	2400	2352	-1	2095	-1	2108	-1	
2258	14	1	2375	2097	-1	2402	2096	-1	-1	
2259	14	2	2376	2403	2404	2132	2098	-1	-1	2097(4)
2260	14	2	2377	2098	2405	-1	2403	2099	-1	
2261	14	2	2401	2405	2098	2102	2404	2100	-1	
2262	14	2	2406	2100	2099	2407	-1	2405	2113	
2263	14	2	2100	2406	2368	2104	-1	2401	2114	
2264	14	2	2399	2102	2101	2408	2128	2103	-1	
2265	14	2	2103	2104	2107	2409	2105	2399	2115	
2266	14	2	2409	2407	-1	2103	2410	2408	2116	
2267	14	2	2106	-1	2117	2105	2411	2112	2117	
2268	14	2	2411	-1	2109	2410	2106	2412	2118	
2269	14	2	2108	2355	2413	2107	2119	2400	2119	
2270	14	2	2120	-1	2412	-1	2125	2109	2414	
2271	14	3	2394	-1	2395	-1	2395	2110	2415	
2272	14	2	2135	-1	2123	2396	2123	2111	2416	

Parameter	ℓ	d	Simples 1	2	3	4	5	6	7	Other
2273	14	2	2124	-1	2121	2397	2112	2417	2111	
2274	14	2	2113	2114	2369	2418	-1	-1	2406	
2275	14	2	2418	2116	-1	2113	2419	-1	2407	
2276	14	2	2116	2418	2420	2115	2421	-1	2409	
2277	14	2	2421	2419	-1	2118	2116	2422	2410	
2278	14	2	2118	-1	2411	2421	2117	2423	2411	
2279	14	2	2423	-1	2120	2422	2122	2118	2424	
2280	14	2	-1	2359	2119	2420	2413	-1	2413	
2281	14	2	2414	-1	2424	-1	2126	2414	2120	
2282	14	2	2122	-1	2417	-1	2423	2121	2425	
2283	14	2	2425	-1	2126	-1	2424	2137	2122	
2284	14	2	2137	-1	2416	-1	2416	2425	2123	
2285	14	2	2397	2426	2138	2124	2127	2427	2134	
2286	14	2	2417	-1	2125	2427	2412	2124	2428	
2287	14	2	2126	-1	2428	2140	2414	2136	2125	
2288	14	2	2388	2127	2141	2129	2426	2429	2133	
2289	14	2	2398	2429	2142	2128	2430	2127	-1	
2290	14	2	2389	2130	2143	2429	2431	2129	-1	2131(3)
2291	14	2	2391	2431	2144	2404	2130	-1	-1	2132(3)
2292	14	1	2390	2132	2431	2432	2131	-1	-1	
2293	14	2	2387	2134	2147	-1	2133	2433	2426	
2294	14	2	2396	2433	2145	2135	-1	2134	2434	
2295	14	2	2416	-1	2136	2434	2137	2428	2135	
2296	14	2	-1	2138	2426	2148	2141	2435	2147	
2297	14	2	-1	2435	2427	2139	2436	2138	2437	
2298	14	2	-1	2150	2140	2437	-1	2146	2139	
2299	14	2	-1	2142	2429	2151	2438	2141	-1	2143(4)
2300	14	2	-1	2438	2430	-1	2142	2436	-1	
2301	14	1	-1	2144	2431	2439	2143	-1	-1	
2302	14	2	-1	2145	2433	2156	-1	2147	2440	
2303	14	2	-1	2440	2434	2146	-1	2437	2145	
2304	14	2	-1	2149	-1	2435	2441	2148	2442	
2305	14	2	-1	2441	-1	-1	2149	2154	2443	2152(6)
2306	14	2	-1	2442	-1	2150	2443	2157	2149	
2307	14	1	-1	2152	-1	2444	2151	2441	-1	
2308	14	1	-1	2154	-1	2445	2153	2441	2446	
2309	14	2	-1	2155	-1	-1	2446	2158	2159	2153(7)
2310	14	2	-1	2446	-1	-1	2155	2443	2160	2154(7)
2311	14	2	-1	2157	-1	2440	2158	2442	2156	
2312	14	1	-1	2160	-1	2447	2159	-1	2446	
2313	15	1	-1	-1	2448	2161	-1	-1	2449	
2314	15	2	-1	2163	-1	2449	2162	2450	2186	2161(7)
2315	15	2	-1	2450	-1	-1	2166	2163	2185	
2316	15	2	2451	2167	2164	-1	-1	2176	2452	
2317	15	2	-1	2168	2452	2165	-1	2453	2164	
2318	15	2	-1	2166	-1	2453	2450	2165	2184	
2319	15	2	2167	2451	-1	2170	-1	2178	2454	
2320	15	2	2454	2452	2168	2171	-1	2455	2167	
2321	15	2	-1	2453	2455	2169	2456	2168	2180	
2322	15	2	2171	-1	2173	2454	2173	2457	2170	
2323	15	2	2457	-1	2172	2455	2458	2171	2182	
2324	15	2	2175	-1	2458	-1	2172	2175	2183	
2325	15	2	2174	-1	2457	-1	2459	2173	2189	
2326	15	2	2459	-1	2175	2460	2174	2458	2190	
2327	15	2	2176	2178	-1	-1	2193	2451	2461	
2328	15	2	2461	2179	2177	-1	2194	2462	2176	
2329	15	2	-1	2180	2462	2184	2463	2177	2453	
2330	15	2	2179	2461	-1	2181	2196	2464	2178	
2331	15	2	2464	2462	2180	2182	2465	2179	2455	
2332	15	2	-1	2463	2465	-1	2180	2198	2456	
2333	15	2	2182	-1	2189	2464	2466	2181	2457	
2334	15	2	2466	-1	2183	2467	2182	2188	2458	
2335	15	2	-1	2185	-1	2468	2184	2201	2450	2186(6)
2336	15	1	-1	-1	2469	2186	-1	2468	2449	
2337	15	2	2188	-1	2191	2470	2187	2466	2191	
2338	15	2	2470	2471	-1	2188	2205	2467	2192	
2339	15	2	2190	-1	2466	2472	2189	2191	2459	
2340	15	2	2472	2473	-1	2190	-1	2192	2460	
2341	15	2	2192	2474	2475	2191	2208	2472	2470	
2342	15	2	2474	2192	-1	-1	2209	2473	2471	
2343	15	2	2194	2196	-1	2212	2461	2476	2193	
2344	15	2	2476	2197	2195	2213	2477	2194	-1	
2345	15	2	-1	2198	2477	2478	2195	2463	-1	
2346	15	2	2197	2476	-1	2199	2479	2196	-1	
2347	15	2	2479	2477	2198	2480	2197	2465	-1	
2348	15	2	2200	2217	2202	2481	2199	2204	-1	

Parameter	ℓ	d	Simples 1	2	3	4	5	6	7	Other
2349	15	2	2481	2482	-1	2200	2480	2205	-1	
2350	15	1	-1	-1	2483	2201	2478	2468	-1	
2351	15	2	2203	2484	2485	2202	-1	2206	-1	
2352	15	3	2484	2257	-1	2224	-1	2207	-1	2203(2)
2353	15	2	2205	2486	2487	2204	2470	2481	2208	
2354	15	2	2486	2205	-1	2220	2471	2482	2209	
2355	15	3	2207	2269	2488	2225	2210	2484	2210	2206(2)
2356	15	2	-1	2488	2206	2487	2211	2485	2211	
2357	15	2	2209	2208	2489	2221	2474	-1	2486	
2358	15	2	-1	2489	2208	2211	2475	-1	2487	
2359	15	3	-1	2280	2210	2490	2488	-1	2488	2211(2)
2360	15	2	2213	2215	2229	2476	2491	2212	-1	
2361	15	2	2491	2216	2235	2492	2213	-1	-1	2214(3)
2362	15	1	-1	-1	2493	2214	2478	-1	-1	
2363	15	2	2216	2491	2230	2494	2215	-1	-1	
2364	15	2	2494	2495	2236	2216	2218	-1	-1	
2365	15	2	2218	2496	2497	2217	2494	2219	-1	
2366	15	2	2496	2218	2237	2482	2495	2220	-1	
2367	15	2	2220	2219	2498	2486	-1	2496	2221	
2368	15	3	2233	2498	2263	2226	-1	2497	2222	2219(3)
2369	15	3	2234	2222	2274	2499	-1	-1	2498	2221(3)
2370	15	3	2224	2223	2500	2484	2239	2225	-1	
2371	15	3	2240	2500	2223	2497	2240	2226	-1	
2372	15	3	2241	2226	2225	2501	2227	2500	2228	
2373	15	3	2249	-1	2238	2227	2502	2249	2238	
2374	15	3	2242	2499	2490	2228	2503	-1	2501	
2375	15	2	2258	2230	2491	2504	2229	-1	-1	2235(1)
2376	15	3	2259	2505	2506	2247	2231	-1	-1	2236(1) 2230(4)
2377	15	3	2260	2231	2507	-1	2505	2232	-1	2237(1)
2378	15	3	2497	2507	2231	2240	2506	2233	-1	
2379	15	3	2498	2233	2232	2508	-1	2507	2234	
2380	15	3	2499	2242	-1	2234	2509	-1	2508	
2381	15	1	2504	2510	2511	2235	-1	-1	-1	
2382	15	2	2505	2236	2495	2510	2237	-1	-1	
2383	15	3	2251	-1	2502	2503	2238	2512	2502	
2384	15	3	2500	2240	2239	2513	2500	2241	-1	
2385	15	3	2501	2508	-1	2241	2514	2513	2242	
2386	15	3	2503	2509	-1	2251	2242	2515	2514	
2387	15	2	2293	2252	-1	-1	2243	2516	2517	
2388	15	2	2288	2253	-1	2244	2517	2518	2243	
2389	15	2	2290	2246	2245	2518	2519	2244	-1	
2390	15	2	2292	2247	2519	2520	2245	-1	-1	2248(1)
2391	15	2	2291	2519	2247	2521	2246	-1	-1	
2392	15	1	2520	2510	2522	2248	-1	-1	-1	
2393	15	3	2502	-1	2250	2514	2249	2523	2251	
2394	15	3	2271	-1	2523	-1	2512	2250	2524	
2395	15	3	2512	-1	2271	2515	2271	2251	2525	
2396	15	2	2294	2516	-1	2272	-1	2252	2526	
2397	15	2	2285	2517	-1	2273	2253	2527	2252	
2398	15	2	2289	2518	-1	2254	2528	2253	-1	
2399	15	2	2264	2256	2255	2529	2254	2265	-1	
2400	15	2	2257	2484	2530	2255	-1	2269	-1	
2401	15	2	2261	2531	2497	2256	2521	2263	-1	
2402	15	1	2504	2532	2533	2258	-1	-1	-1	
2403	15	2	2505	2259	2534	2532	2260	-1	-1	
2404	15	2	2521	2534	2259	2291	2261	-1	-1	
2405	15	2	2531	2261	2260	2535	2534	2262	-1	
2406	15	2	2262	2263	2498	2536	-1	2531	2274	
2407	15	2	2536	2266	-1	2262	2537	2535	2275	
2408	15	2	2529	2535	-1	2264	2538	2266	-1	
2409	15	2	2266	2536	2539	2265	2540	2529	2276	
2410	15	2	2540	2537	-1	2268	2266	2541	2277	
2411	15	2	2268	-1	2278	2540	2267	2542	2278	
2412	15	2	2542	-1	2270	2541	2286	2268	2543	
2413	15	2	-1	2488	2269	2539	2280	2530	2280	
2414	15	2	2281	-1	2543	-1	2287	2281	2270	
2415	15	3	2524	-1	2525	-1	2525	2524	2271	
2416	15	2	2295	-1	2284	2526	2284	2544	2272	
2417	15	2	2286	-1	2282	2527	2542	2273	2544	
2418	15	2	2275	2276	2545	2274	2546	-1	2536	
2419	15	2	2546	2277	-1	-1	2275	2547	2537	
2420	15	2	-1	2545	2276	2280	2548	-1	2539	
2421	15	2	2277	2546	2548	2278	2276	2549	2540	
2422	15	2	2549	2547	-1	2279	-1	2277	2550	
2423	15	2	2279	-1	2542	2549	2282	2278	2551	
2424	15	2	2551	-1	2281	2550	2283	2543	2279	

Para-meter	ℓ	d	Simples 1	2	3	4	5	6	7	Other
2425	15	2	2283	-1	2544	-1	2551	2284	2282	
2426	15	2	2517	2285	2296	-1	2288	2552	2293	
2427	15	2	2527	2552	2297	2286	2553	2285	2554	
2428	15	2	2544	-1	2287	2554	2543	2295	2286	
2429	15	2	2518	2289	2299	2290	2555	2288	-1	
2430	15	2	2528	2555	2300	2538	2289	2553	-1	
2431	15	2	2519	2291	2301	2556	2290	-1	-1	2292(3)
2432	15	1	2520	2532	2557	2292	-1	-1	-1	
2433	15	2	2516	2294	2302	-1	-1	2293	2558	
2434	15	2	2526	2558	2303	2295	-1	2554	2294	
2435	15	2	-1	2297	2552	2304	2559	2296	2560	
2436	15	2	-1	2559	2553	-1	2297	2300	2561	
2437	15	2	-1	2560	2554	2298	2561	2303	2297	
2438	15	2	-1	2300	2555	2562	2299	2559	-1	
2439	15	1	-1	-1	2563	2301	2562	-1	-1	
2440	15	2	-1	2303	2558	2311	-1	2560	2302	
2441	15	2	-1	2305	-1	2564	2304	2308	2565	2307(6)
2442	15	2	-1	2306	-1	2560	2565	2311	2304	
2443	15	2	-1	2565	-1	-1	2306	2310	2305	
2444	15	1	-1	-1	2566	2307	2562	2564	-1	
2445	15	1	-1	-1	2567	2308	-1	2564	2568	
2446	15	2	-1	2310	-1	2568	2309	2565	2312	2308(7)
2447	15	1	-1	-1	2569	2312	-1	-1	2568	
2448	16	1	2570	-1	2313	-1	-1	-1	2571	
2449	16	2	-1	-1	2571	2314	-1	2572	2336	2313(7)
2450	16	2	-1	2315	-1	2572	2318	2314	2335	
2451	16	2	2316	2319	-1	-1	-1	2327	2573	
2452	16	2	2573	2320	2317	-1	-1	2574	2316	
2453	16	2	-1	2321	2574	2318	2575	2317	2329	
2454	16	2	2320	2573	-1	2322	-1	2576	2319	
2455	16	2	2576	2574	2321	2323	2577	2320	2331	
2456	16	2	-1	2575	2577	-1	2321	-1	2332	
2457	16	2	2323	-1	2325	2576	2578	2322	2333	
2458	16	2	2578	-1	2324	2579	2323	2326	2334	
2459	16	2	2326	-1	2578	2580	2325	2578	2339	
2460	16	2	2580	2581	-1	2326	-1	2579	2340	
2461	16	2	2328	2330	-1	-1	2343	2582	2327	
2462	16	2	2582	2331	2329	-1	2583	2328	2574	
2463	16	2	-1	2332	2583	2584	2329	2345	2575	
2464	16	2	2331	2582	-1	2333	2585	2330	2576	
2465	16	2	2585	2583	2332	2586	2331	2347	2577	
2466	16	2	2334	-1	2339	2587	2333	2337	2578	
2467	16	2	2587	2588	-1	2334	2586	2338	2579	
2468	16	2	-1	-1	2589	2335	2584	2350	2572	2336(6)
2469	16	1	2590	-1	2336	-1	-1	2589	2571	
2470	16	2	2338	2591	2592	2337	2353	2587	2341	
2471	16	2	2591	2338	-1	-1	2354	2588	2342	
2472	16	2	2340	2593	2594	2339	-1	2341	2580	
2473	16	2	2593	2340	-1	-1	-1	2342	2581	
2474	16	2	2342	2341	2595	-1	2357	2593	2591	
2475	16	2	-1	2595	2341	2592	2358	2594	2592	
2476	16	2	2344	2346	-1	2360	2596	2343	-1	
2477	16	2	2596	2347	2345	2597	2344	2583	-1	
2478	16	2	-1	-1	2598	2345	2362	2584	-1	2350(5)
2479	16	2	2347	2596	-1	2599	2346	2585	-1	
2480	16	2	2599	2600	-1	2347	2349	2586	-1	
2481	16	2	2349	2601	2602	2348	2599	2353	-1	
2482	16	2	2601	2349	-1	2366	2600	2354	-1	
2483	16	1	2603	-1	2350	-1	2598	2589	-1	
2484	16	3	2352	2400	2604	2370	-1	2355	-1	2351(2)
2485	16	2	-1	2604	2351	2602	-1	2356	-1	
2486	16	2	2354	2353	2605	2367	2591	2601	2357	
2487	16	2	-1	2605	2353	2356	2592	2602	2358	
2488	16	3	-1	2413	2355	2606	2359	2604	2359	2356(2)
2489	16	2	-1	2358	2357	2607	2595	-1	2605	
2490	16	3	-1	2607	2374	2359	2608	-1	2606	
2491	16	2	2361	2363	2375	2609	2360	-1	-1	
2492	16	2	2609	2610	2611	2361	2597	-1	-1	
2493	16	1	2612	-1	2362	2611	2598	-1	-1	
2494	16	2	2364	2613	2614	2363	2365	-1	-1	
2495	16	2	2613	2364	2382	2610	2366	-1	-1	
2496	16	2	2366	2365	2615	2601	2613	2367	-1	
2497	16	3	2378	2615	2401	2371	2614	2368	-1	2365(3)
2498	16	3	2379	2368	2406	2616	-1	2615	2369	2367(3)
2499	16	3	2380	2374	2607	2369	2617	-1	2616	
2500	16	3	2384	2371	2370	2618	2384	2372	-1	
2501	16	3	2385	2616	2606	2372	2619	2618	2374	
2502	16	3	2393	-1	2383	2619	2373	2620	2383	
2503	16	3	2386	2617	2608	2383	2374	2621	2619	
2504	16	2	2402	2622	2623	2375	-1	-1	-1	2381(1)
2505	16	3	2403	2376	2624	2622	2377	-1	-1	2382(1)
2506	16	3	2614	2624	2376	2614	2378	-1	-1	
2507	16	3	2615	2378	2377	2625	2624	2379	-1	
2508	16	3	2616	2385	-1	2379	2626	2625	2380	
2509	16	3	2617	2386	-1	-1	2380	2627	2626	
2510	16	2	2622	2392	2628	2382	-1	-1	-1	2381(2)
2511	16	1	2629	2628	2381	2611	-1	-1	-1	
2512	16	3	2395	-1	2620	2621	2394	2383	2630	
2513	16	3	2618	2625	-1	2384	2631	2385	-1	
2514	16	3	2619	2626	-1	2393	2385	2632	2386	
2515	16	3	2621	2627	-1	2395	-1	2386	2633	
2516	16	2	2433	2396	-1	-1	-1	2387	2634	
2517	16	2	2426	2397	-1	-1	2388	2635	2387	
2518	16	2	2429	2398	-1	2389	2636	2388	-1	
2519	16	2	2431	2391	2390	2637	2389	-1	-1	
2520	16	2	2432	2622	2638	2390	-1	-1	-1	2392(1)
2521	16	2	2404	2639	2614	2391	2401	-1	-1	
2522	16	1	2640	2628	2392	-1	-1	-1	-1	
2523	16	3	2620	-1	2394	2632	2620	2393	2641	
2524	16	3	2415	-1	2641	-1	2630	2415	2394	
2525	16	3	2630	-1	2415	2633	2415	2641	2395	
2526	16	2	2434	2634	-1	2416	-1	2642	2396	
2527	16	2	2427	2635	-1	2417	2643	2397	2642	
2528	16	2	2430	2636	-1	2644	2398	2643	-1	
2529	16	2	2408	2645	2646	2399	2644	2409	-1	
2530	16	2	-1	2604	2400	2646	-1	2413	-1	
2531	16	2	2405	2401	2615	2645	2639	2406	-1	
2532	16	2	2622	2432	2647	2403	-1	-1	-1	2402(2)
2533	16	1	2648	2647	2402	-1	-1	-1	-1	
2534	16	2	2639	2404	2403	2649	2405	-1	-1	
2535	16	2	2645	2408	-1	2405	2650	2407	-1	
2536	16	2	2407	2409	2651	2406	2652	2645	2418	
2537	16	2	2652	2410	-1	-1	2407	2653	2419	
2538	16	2	2644	2650	-1	2430	2408	2654	-1	
2539	16	2	-1	2651	2409	2413	2655	2646	2420	
2540	16	2	2410	2652	2655	2411	2409	2656	2421	
2541	16	2	2656	2653	-1	2412	2654	2410	2657	
2542	16	2	2412	-1	2423	2656	2417	2411	2658	
2543	16	2	2658	-1	2414	2657	2428	2424	2412	
2544	16	2	2428	-1	2425	2642	2658	2416	2417	
2545	16	2	-1	2420	2418	2607	2659	-1	2651	
2546	16	2	2419	2421	2659	-1	2418	2660	2652	
2547	16	2	2660	2422	-1	-1	-1	2419	2661	
2548	16	2	-1	2659	2421	2655	2420	2662	2655	
2549	16	2	2422	2660	2662	2423	-1	2421	2663	
2550	16	2	2663	2661	-1	2424	-1	2657	2422	
2551	16	2	2424	-1	2658	2663	2425	2658	2423	
2552	16	2	2635	2427	2435	-1	2664	2426	2665	
2553	16	2	2643	2664	2436	2654	2427	2430	2666	
2554	16	2	2642	2665	2437	2428	2666	2434	2427	
2555	16	2	2636	2430	2438	2667	2429	2664	-1	
2556	16	2	2637	2649	2668	2431	2667	-1	-1	
2557	16	1	2669	2647	2432	2668	-1	-1	-1	
2558	16	2	2634	2434	2440	-1	-1	2665	2433	
2559	16	2	-1	2436	2664	2670	2435	2438	2671	
2560	16	2	-1	2437	2665	2442	2671	2440	2435	
2561	16	2	-1	2671	2666	-1	2437	-1	2436	
2562	16	2	-1	-1	2672	2438	2444	2670	-1	2439(5)
2563	16	1	2673	-1	2439	2668	2672	-1	-1	
2564	16	2	-1	-1	2674	2441	2670	2445	2675	2444(6)
2565	16	2	-1	2443	-1	2675	2442	2446	2441	
2566	16	1	2676	-1	2444	-1	2672	2674	-1	
2567	16	1	2677	-1	2445	-1	-1	2674	2678	
2568	16	2	-1	-1	2678	2446	-1	2675	2447	2445(7)
2569	16	1	2679	-1	2447	-1	-1	-1	2678	
2570	17	1	2448	-1	-1	-1	-1	-1	2680	
2571	17	2	2680	-1	2449	-1	-1	2681	2469	2448(7)
2572	17	2	-1	-1	2681	2450	2682	2449	2468	
2573	17	2	2452	2454	-1	-1	-1	2683	2451	
2574	17	2	2683	2455	2453	-1	2684	2452	2462	
2575	17	2	-1	2456	2684	2682	2453	-1	2463	
2576	17	2	2455	2683	-1	2457	2685	2454	2464	

Parameter	ℓ	d	Simples 1	2	3	4	5	6	7	Other
2577	17	2	2685	2684	2456	2686	2455	-1	2465	
2578	17	2	2458	-1	2459	2687	2457	2459	2466	
2579	17	2	2687	2688	-1	2458	2686	2460	2467	
2580	17	2	2460	2689	2690	2459	-1	2687	2472	
2581	17	2	2689	2460	-1	-1	-1	2688	2473	
2582	17	2	2462	2464	-1	-1	2691	2461	2683	
2583	17	2	2691	2465	2463	2692	2462	2477	2684	
2584	17	2	-1	-1	2693	2463	2468	2478	2682	
2585	17	2	2465	2691	-1	2694	2464	2479	2685	
2586	17	2	2694	2695	-1	2465	2467	2480	2686	
2587	17	2	2467	2696	2697	2466	2694	2470	2687	
2588	17	2	2696	2467	-1	-1	2695	2471	2688	
2589	17	2	2698	-1	2468	-1	2693	2483	2681	2469(6)
2590	17	1	2469	-1	-1	-1	-1	2698	2680	
2591	17	2	2471	2470	2699	-1	2486	2696	2474	
2592	17	2	-1	2699	2470	2475	2487	2697	2475	
2593	17	2	2473	2472	2700	-1	-1	2474	2689	
2594	17	2	-1	2700	2472	2697	-1	2475	2690	
2595	17	2	-1	2475	2474	2701	2489	2700	2699	
2596	17	2	2477	2479	-1	2702	2476	2691	-1	
2597	17	2	2702	2703	2704	2477	2492	2692	-1	
2598	17	2	2705	-1	2478	2704	2493	2693	-1	2483(5)
2599	17	2	2480	2706	2707	2479	2481	2694	-1	
2600	17	2	2706	2480	-1	2703	2482	2695	-1	
2601	17	2	2482	2481	2708	2496	2706	2486	-1	
2602	17	2	-1	2708	2481	2485	2707	2487	-1	
2603	17	1	2483	-1	-1	-1	2705	2698	-1	
2604	17	3	-1	2530	2484	2709	-1	2488	-1	2485(2)
2605	17	2	-1	2487	2486	2710	2699	2708	2489	
2606	17	3	-1	2710	2501	2488	2711	2709	2490	
2607	17	3	-1	2490	2499	2545	2712	-1	2710	2489(4)
2608	17	3	-1	2712	2503	2711	2490	2713	2711	
2609	17	2	2492	2714	2715	2491	2702	-1	-1	
2610	17	2	2714	2492	2716	2495	2703	-1	-1	
2611	17	2	2717	2716	2492	2511	2704	-1	-1	2493(4)
2612	17	1	2493	-1	-1	2717	2705	-1	-1	
2613	17	2	2495	2494	2718	2714	2496	-1	-1	
2614	17	3	2506	2718	2521	2506	2497	-1	-1	2494(3)
2615	17	3	2507	2497	2531	2719	2718	2498	-1	2496(3)
2616	17	3	2508	2501	2710	2498	2720	2719	2499	
2617	17	3	2509	2503	2712	-1	2499	2721	2720	
2618	17	3	2513	2719	2709	2500	2722	2501	-1	
2619	17	3	2514	2720	2711	2502	2501	2723	2503	
2620	17	3	2523	-1	2512	2723	2523	2502	2724	
2621	17	3	2515	2721	2713	2512	-1	2503	2725	
2622	17	3	2532	2520	2726	2505	-1	-1	-1	2510(1) 2504(2)
2623	17	2	2727	2726	2504	2715	-1	-1	-1	
2624	17	3	2718	2506	2505	2728	2507	-1	-1	
2625	17	3	2719	2513	-1	2507	2729	2508	-1	
2626	17	3	2720	2514	-1	-1	2508	2730	2509	
2627	17	3	2721	2515	-1	-1	-1	2509	2731	
2628	17	2	2732	2522	2510	2716	-1	-1	-1	2511(2)
2629	17	1	2511	2732	2727	2717	-1	-1	-1	
2630	17	3	2525	-1	2724	2725	2524	2724	2512	
2631	17	3	2722	2729	-1	2722	2513	2733	-1	
2632	17	3	2723	2730	-1	2523	2733	2514	2734	
2633	17	3	2725	2731	-1	2525	-1	2734	2515	
2634	17	2	2558	2526	-1	-1	-1	2735	2516	
2635	17	2	2552	2527	-1	-1	2736	2517	2735	
2636	17	2	2555	2528	-1	2737	2518	2736	-1	
2637	17	2	2556	2738	2739	2519	2737	-1	-1	
2638	17	2	2740	2726	2520	2739	-1	-1	-1	
2639	17	2	2534	2521	2718	2738	2531	-1	-1	
2640	17	1	2522	2732	2740	-1	-1	-1	-1	
2641	17	3	2724	-1	2524	2734	2724	2525	2523	
2642	17	2	2554	2735	-1	2544	2741	2526	2527	
2643	17	2	2553	2736	-1	2742	2527	2528	2741	
2644	17	2	2538	2743	2744	2528	2529	2742	-1	
2645	17	2	2535	2529	2745	2531	2743	2536	-1	
2646	17	2	-1	2745	2529	2530	2744	2539	-1	
2647	17	2	2746	2557	2532	2747	-1	-1	-1	2533(2)
2648	17	1	2533	2746	2727	-1	-1	-1	-1	
2649	17	2	2738	2556	2747	2534	2748	-1	-1	
2650	17	2	2743	2538	-1	2748	2535	2749	-1	
2651	17	2	-1	2539	2536	2710	2750	2745	2545	
2652	17	2	2537	2540	2750	-1	2536	2751	2546	
2653	17	2	2751	2541	-1	-1	2749	2537	2752	
2654	17	2	2742	2749	-1	2553	2541	2538	2753	
2655	17	2	-1	2750	2540	2548	2539	2754	2548	
2656	17	2	2541	2751	2754	2542	2742	2540	2755	
2657	17	2	2755	2752	-1	2543	2753	2550	2541	
2658	17	2	2543	-1	2551	2755	2544	2551	2542	
2659	17	2	-1	2548	2546	2756	2545	2757	2750	
2660	17	2	2547	2549	2757	-1	-1	2546	2758	
2661	17	2	2758	2550	-1	-1	-1	2752	2547	
2662	17	2	-1	2757	2549	2754	-1	2548	2759	
2663	17	2	2550	2758	2759	2551	-1	2755	2549	
2664	17	2	2736	2553	2559	2760	2552	2555	2761	
2665	17	2	2735	2554	2560	-1	2761	2558	2552	
2666	17	2	2741	2761	2561	2753	2554	-1	2553	
2667	17	2	2737	2748	2762	2555	2556	2760	-1	
2668	17	2	2763	2747	2556	2563	2762	-1	-1	2557(4)
2669	17	1	2557	2746	2740	2763	-1	-1	-1	
2670	17	2	-1	-1	2764	2559	2564	2562	2765	
2671	17	2	-1	2561	2761	2765	2560	-1	2559	
2672	17	2	2766	-1	2562	2762	2566	2764	-1	2563(5)
2673	17	1	2563	-1	-1	2763	2766	-1	-1	
2674	17	2	2767	-1	2564	-1	2764	2567	2768	2566(6)
2675	17	2	-1	-1	2768	2565	2765	2568	2564	
2676	17	1	2566	-1	-1	-1	2766	2767	-1	
2677	17	1	2567	-1	-1	-1	-1	2767	2769	
2678	17	2	2769	-1	2568	-1	-1	2768	2569	2567(7)
2679	17	1	2569	-1	-1	-1	-1	-1	2769	
2680	18	2	2571	-1	-1	-1	-1	2770	2590	2570(7)
2681	18	2	2770	-1	2572	-1	2771	2571	2589	
2682	18	2	-1	-1	2771	2575	2572	-1	2584	
2683	18	2	2574	2576	-1	-1	2772	2573	2582	
2684	18	2	2772	2577	2575	2773	2574	-1	2583	
2685	18	2	2577	2772	-1	2774	2576	-1	2585	
2686	18	2	2774	2775	-1	2577	2579	-1	2586	
2687	18	2	2579	2776	2777	2578	2774	2580	2587	
2688	18	2	2776	2579	-1	-1	2775	2581	2588	
2689	18	2	2581	2580	2778	-1	-1	2776	2593	
2690	18	2	-1	2778	2580	2777	-1	2777	2594	
2691	18	2	2583	2585	-1	2779	2582	2596	2772	
2692	18	2	2779	2780	2781	2583	-1	2597	2773	
2693	18	2	2782	-1	2584	2781	2589	2598	2771	
2694	18	2	2586	2783	2784	2585	2587	2599	2774	
2695	18	2	2783	2586	-1	2780	2588	2600	2775	
2696	18	2	2588	2587	2785	-1	2783	2591	2776	
2697	18	2	-1	2785	2587	2594	2784	2592	2777	
2698	18	2	2589	-1	-1	-1	2782	2603	2770	2590(6)
2699	18	2	-1	2592	2591	2786	2605	2785	2595	
2700	18	2	-1	2594	2593	2787	-1	2595	2778	
2701	18	2	-1	2786	-1	2595	2788	2787	2786	
2702	18	2	2597	2789	2790	2596	2609	2779	-1	
2703	18	2	2789	2597	2791	2600	2610	2780	-1	
2704	18	2	2792	2791	2597	2598	2611	2781	-1	
2705	18	2	2598	-1	-1	2792	2612	2782	-1	2603(5)
2706	18	2	2600	2599	2793	2789	2601	2783	-1	
2707	18	2	-1	2793	2599	-1	2602	2784	-1	
2708	18	2	-1	2602	2601	2794	2793	2605	-1	
2709	18	3	-1	2794	2618	2604	2795	2606	-1	
2710	18	3	-1	2606	2616	2651	2796	2794	2607	2605(4)
2711	18	3	-1	2796	2619	2608	2606	2797	2608	
2712	18	3	-1	2608	2617	2788	2607	2798	2796	
2713	18	3	-1	2798	2621	2797	-1	2608	2799	
2714	18	2	2610	2609	2800	2613	2789	-1	-1	
2715	18	2	2801	2800	2609	2623	2790	-1	-1	
2716	18	2	2802	2611	2610	2628	2791	-1	-1	
2717	18	2	2611	2802	2801	2629	2792	-1	-1	2612(4)
2718	18	3	2624	2614	2639	2803	2615	-1	-1	2613(3)
2719	18	3	2625	2618	2794	2615	2804	2616	-1	
2720	18	3	2626	2619	2796	-1	2616	2805	2617	
2721	18	3	2627	2621	2798	-1	-1	2617	2806	
2722	18	3	2631	2804	2795	2631	2618	2807	-1	
2723	18	3	2632	2805	2797	2620	2807	2619	2808	
2724	18	3	2641	-1	2630	2808	2641	2630	2620	
2725	18	3	2633	2806	2799	2630	-1	2808	2621	
2726	18	3	2809	2638	2622	2810	-1	-1	-1	2623(2)
2727	18	2	2623	2809	2648	2801	-1	-1	-1	2629(3)
2728	18	3	2803	2803	2810	2624	2811	-1	-1	

Parameter	ℓ	d	Simples 1	2	3	4	5	6	7	Other
2729	18	3	2804	2631	-1	2811	2625	2812	-1	
2730	18	3	2805	2632	-1	-1	2812	2626	2813	
2731	18	3	2806	2633	-1	-1	-1	2813	2627	
2732	18	2	2628	2640	2809	2802	-1	-1	-1	2629(2)
2733	18	3	2807	2812	-1	2807	2632	2631	2814	
2734	18	3	2808	2813	-1	2641	2814	2633	2632	
2735	18	2	2665	2642	-1	-1	2815	2634	2635	
2736	18	2	2664	2643	-1	2816	2635	2636	2815	
2737	18	2	2667	2817	2818	2636	2637	2816	-1	
2738	18	2	2649	2637	2819	2639	2817	-1	-1	
2739	18	2	2820	2819	2637	2638	2818	-1	-1	
2740	18	2	2638	2809	2669	2820	-1	-1	-1	2640(3)
2741	18	2	2666	2815	-1	2821	2642	-1	2643	
2742	18	2	2654	2822	2823	2643	2656	2644	2821	
2743	18	2	2650	2644	2824	2817	2645	2822	-1	
2744	18	2	-1	2824	2644	-1	2646	2823	-1	
2745	18	2	-1	2646	2645	2794	2824	2651	-1	
2746	18	2	2647	2669	2809	2825	-1	-1	-1	2648(2)
2747	18	2	2825	2668	2649	2647	2826	-1	-1	
2748	18	2	2817	2667	2826	2650	2649	2827	-1	
2749	18	2	2822	2654	-1	2827	2653	2650	2828	
2750	18	2	-1	2655	2652	2829	2651	2830	2659	
2751	18	2	2653	2656	2830	-1	2822	2652	2831	
2752	18	2	2831	2657	-1	-1	2828	2661	2653	
2753	18	2	2821	2828	-1	2666	2657	-1	2654	
2754	18	2	-1	2830	2656	2662	2823	2655	2832	
2755	18	2	2657	2831	2832	2658	2821	2663	2656	
2756	18	2	-1	2829	-1	2659	2788	2833	2829	
2757	18	2	-1	2662	2660	2833	-1	2659	2834	
2758	18	2	2661	2663	2834	-1	-1	2831	2660	
2759	18	2	-1	2834	2663	2832	-1	2832	2662	
2760	18	2	2816	2827	2835	2664	-1	2667	2836	
2761	18	2	2815	2666	2671	2836	2665	-1	2664	
2762	18	2	2837	2826	2667	2672	2668	2835	-1	
2763	18	2	2668	2825	2820	2673	2837	-1	-1	2669(4)
2764	18	2	2838	-1	2670	2835	2674	2672	2839	
2765	18	2	-1	-1	2839	2671	2675	-1	2670	
2766	18	2	2672	-1	-1	2837	2676	2838	-1	2673(5)
2767	18	2	2674	-1	-1	-1	2838	2677	2840	2676(6)
2768	18	2	2840	-1	2675	-1	2839	2678	2674	
2769	18	2	2678	-1	-1	-1	-1	2840	2679	2677(7)
2770	19	2	2681	-1	-1	-1	2841	2680	2698	
2771	19	2	2841	-1	2682	2842	2681	-1	2693	
2772	19	2	2684	2685	-1	2843	2683	-1	2691	
2773	19	2	2843	2844	2842	2684	-1	-1	2692	
2774	19	2	2686	2845	2846	2685	2687	-1	2694	
2775	19	2	2845	2686	-1	2844	2688	-1	2695	
2776	19	2	2688	2687	2847	-1	2845	2689	2696	
2777	19	2	-1	2847	2687	2690	2846	2690	2697	
2778	19	2	-1	2690	2689	2848	-1	2847	2700	
2779	19	2	2692	2849	2850	2691	-1	2702	2843	
2780	19	2	2849	2692	2851	2695	-1	2703	2844	
2781	19	2	2852	2851	2692	2693	-1	2704	2842	
2782	19	2	2693	-1	-1	2852	2698	2705	2841	
2783	19	2	2695	2694	2853	2849	2696	2706	2845	
2784	19	2	-1	2853	2694	-1	2697	2707	2846	
2785	19	2	-1	2697	2696	2854	2853	2699	2847	
2786	19	2	-1	2701	-1	2699	2855	2854	2701	
2787	19	2	-1	2854	-1	2700	2856	2701	2848	
2788	19	3	-1	2855	-1	2712	2756	2857	2855	2701(5)
2789	19	2	2703	2702	2858	2706	2714	2849	-1	
2790	19	2	2859	2858	2702	-1	2715	2850	-1	
2791	19	2	2860	2704	2703	-1	2716	2851	-1	
2792	19	2	2704	2860	2859	2705	2717	2852	-1	
2793	19	2	-1	2707	2706	2861	2708	2853	-1	
2794	19	3	-1	2709	2719	2745	2862	2710	-1	2708(4)
2795	19	3	-1	2862	2722	-1	2709	2863	-1	
2796	19	3	-1	2711	2720	2855	2710	2864	2712	
2797	19	3	-1	2864	2723	2713	2863	2711	2865	
2798	19	3	-1	2713	2721	2857	-1	2712	2866	
2799	19	3	-1	2866	2725	2865	-1	2865	2713	
2800	19	2	2867	2715	2714	2868	2858	-1	-1	
2801	19	2	2715	2867	2717	2727	2859	-1	-1	
2802	19	2	2716	2717	2867	2732	2860	-1	-1	
2803	19	3	2728	2728	2868	2718	2869	-1	-1	
2804	19	3	2729	2722	2862	2869	2719	2870	-1	
2805	19	3	2730	2723	2864	-1	2870	2720	2871	
2806	19	3	2731	2725	2866	-1	-1	2871	2721	
2807	19	3	2733	2870	2863	2733	2723	2722	2872	
2808	19	3	2734	2871	2865	2724	2872	2725	2723	
2809	19	3	2726	2740	2746	2873	-1	-1	-1	2727(2) 2732(3)
2810	19	3	2873	2868	2728	2726	2874	-1	-1	
2811	19	3	2869	2869	2874	2729	2728	2875	-1	
2812	19	3	2870	2733	-1	2875	2730	2729	2876	
2813	19	3	2871	2734	-1	-1	2876	2731	2730	
2814	19	3	2872	2876	-1	2872	2734	-1	2733	
2815	19	2	2761	2741	-1	2877	2735	-1	2736	
2816	19	2	2760	2878	2879	2736	-1	2737	2877	
2817	19	2	2748	2737	2880	2743	2738	2878	-1	
2818	19	2	2881	2880	2737	-1	2739	2879	-1	
2819	19	2	2882	2739	2738	2868	2880	-1	-1	
2820	19	2	2739	2882	2763	2740	2881	-1	-1	
2821	19	2	2753	2883	2884	2741	2755	-1	2742	
2822	19	2	2749	2742	2885	2878	2751	2743	2883	
2823	19	2	-1	2885	2742	-1	2754	2744	2884	
2824	19	2	-1	2744	2743	2886	2745	2885	-1	
2825	19	2	2747	2763	2882	2746	2887	-1	-1	
2826	19	2	2887	2762	2748	-1	2747	2888	-1	
2827	19	2	2878	2760	2888	2749	-1	2748	2889	
2828	19	2	2883	2753	-1	2889	2752	-1	2749	
2829	19	2	-1	2756	-1	2750	2855	2890	2756	
2830	19	2	-1	2754	2751	2890	2885	2750	2891	
2831	19	2	2752	2755	2891	-1	2883	2758	2751	
2832	19	2	-1	2891	2755	2759	2884	2759	2754	
2833	19	2	-1	2890	-1	2757	2892	2756	2893	
2834	19	2	-1	2759	2758	2893	-1	2891	2757	
2835	19	2	2894	2888	2760	2764	-1	2762	2895	
2836	19	2	2877	2889	2895	2761	-1	-1	2760	
2837	19	2	2762	2887	2881	2766	2763	2894	-1	
2838	19	2	2764	-1	-1	2894	2767	2766	2896	
2839	19	2	2896	-1	2765	2895	2768	-1	2764	
2840	19	2	2768	-1	-1	-1	2896	2769	2767	
2841	20	2	2771	-1	-1	2897	2770	-1	2782	
2842	20	2	2897	2898	2773	2771	-1	-1	2781	
2843	20	2	2773	2899	2900	2772	-1	-1	2779	
2844	20	2	2899	2773	2898	2775	-1	-1	2780	
2845	20	2	2775	2774	2901	2899	2776	-1	2783	
2846	20	2	-1	2901	2774	-1	2777	-1	2784	
2847	20	2	-1	2777	2776	2902	2901	2778	2785	
2848	20	2	-1	2902	-1	2778	2903	2902	2787	
2849	20	2	2780	2779	2904	2783	-1	2789	2899	
2850	20	2	2905	2904	2779	-1	-1	2790	2900	
2851	20	2	2906	2781	2780	-1	-1	2791	2898	
2852	20	2	2781	2906	2905	2782	-1	2792	2897	
2853	20	2	-1	2784	2783	2907	2785	2793	2901	
2854	20	2	-1	2787	-1	2785	2908	2786	2902	
2855	20	3	-1	2788	-1	2796	2829	2909	2788	2786(5)
2856	20	2	-1	2908	-1	-1	2787	2910	2903	
2857	20	3	-1	2909	-1	2798	2910	2788	2911	
2858	20	2	2912	2790	2789	2913	2800	2904	-1	
2859	20	2	2790	2912	2792	-1	2801	2905	-1	
2860	20	2	2791	2792	2912	-1	2802	2906	-1	
2861	20	2	-1	-1	2913	2793	2914	2907	-1	
2862	20	3	-1	2795	2804	2914	2794	2915	-1	
2863	20	3	-1	2915	2807	-1	2797	2795	2916	
2864	20	3	-1	2797	2805	2909	2915	2796	2917	
2865	20	3	-1	2917	2808	2799	2916	2799	2797	
2866	20	3	-1	2799	2806	2911	-1	2917	2798	
2867	20	2	2800	2801	2802	2918	2912	-1	-1	
2868	20	3	2918	2810	2803	2819	2919	-1	-1	2800(4)
2869	20	3	2811	2811	2919	2804	2803	2920	-1	
2870	20	3	2812	2807	2915	2920	2805	2804	2921	
2871	20	3	2813	2808	2917	-1	2921	2806	2805	
2872	20	3	2814	2921	2916	2814	2808	-1	2807	
2873	20	3	2810	2918	2918	2809	2922	-1	-1	
2874	20	3	2922	2919	2811	-1	2810	2923	-1	
2875	20	3	2920	2920	2923	2812	-1	2811	2924	
2876	20	3	2921	2814	-1	2924	2813	-1	2812	
2877	20	2	2836	2925	2926	2815	-1	-1	2816	
2878	20	2	2827	2816	2927	2822	-1	2817	2925	
2879	20	2	2928	2927	2816	-1	-1	2818	2926	
2880	20	2	2929	2818	2817	2930	2819	2927	-1	

Para-meter	ℓ	d	Simples							Other
			1	2	3	4	5	6	7	
2881	20	2	2818	2929	2837	-1	2820	2928	-1	
2882	20	2	2819	2820	2825	2918	2929	-1	-1	
2883	20	2	2828	2821	2931	2925	2831	-1	2822	
2884	20	2	-1	2931	2821	-1	2832	-1	2823	
2885	20	2	-1	2823	2822	2932	2830	2824	2931	
2886	20	2	-1	-1	2930	2824	2914	2932	-1	
2887	20	2	2826	2837	2929	-1	2825	2933	-1	
2888	20	2	2933	2835	2827	-1	-1	2826	2934	
2889	20	2	2925	2836	2934	2828	-1	-1	2827	
2890	20	2	-1	2833	-1	2830	2935	2829	2936	
2891	20	2	-1	2832	2831	2936	2931	2834	2830	
2892	20	2	-1	2935	-1	-1	2833	2910	2937	
2893	20	2	-1	2936	-1	2834	2937	2936	2833	
2894	20	2	2835	2933	2928	2838	-1	2837	2938	
2895	20	2	2938	2934	2836	2839	-1	-1	2835	
2896	20	2	2839	-1	-1	2938	2840	-1	2838	
2897	21	2	2842	2939	2940	2841	-1	-1	2852	
2898	21	2	2939	2842	2844	-1	-1	-1	2851	
2899	21	2	2844	2843	2941	2845	-1	-1	2849	
2900	21	2	2940	2941	2843	-1	-1	-1	2850	
2901	21	2	-1	2846	2845	2942	2847	-1	2853	
2902	21	2	-1	2848	-1	2847	2943	2848	2854	
2903	21	2	-1	2943	-1	-1	2848	2944	2856	
2904	21	2	2945	2850	2849	2946	-1	2858	2941	
2905	21	2	2850	2945	2852	-1	-1	2859	2940	
2906	21	2	2851	2852	2945	-1	-1	2860	2939	
2907	21	2	-1	-1	2946	2853	2947	2861	2942	
2908	21	2	-1	2856	-1	2947	2854	2948	2943	
2909	21	3	-1	2857	-1	2864	2948	2855	2949	
2910	21	3	-1	2948	-1	-1	2857	2892	2950	2856(6)
2911	21	3	-1	2949	-1	2866	2950	2949	2857	
2912	21	2	2858	2859	2860	2951	2867	2945	-1	
2913	21	2	2951	-1	2861	2858	2952	2946	-1	
2914	21	3	-1	-1	2952	2862	2886	2953	-1	2861(5)
2915	21	3	-1	2863	2870	2953	2864	2862	2954	
2916	21	3	-1	2954	2872	-1	2865	-1	2863	
2917	21	3	-1	2865	2871	2949	2954	2866	2864	
2918	21	3	2868	2873	2873	2882	2955	-1	-1	2867(4)
2919	21	3	2955	2874	2869	2952	2868	2956	-1	
2920	21	3	2875	2875	2956	2870	-1	2869	2957	
2921	21	3	2876	2872	2954	2957	2871	-1	2870	
2922	21	3	2874	2955	2955	-1	2873	2958	-1	
2923	21	3	2958	2956	2875	-1	-1	2874	2959	
2924	21	3	2957	2957	2959	2876	-1	-1	2875	
2925	21	2	2889	2877	2960	2883	-1	-1	2878	
2926	21	2	2961	2960	2877	-1	-1	-1	2879	
2927	21	2	2962	2879	2878	2963	-1	2880	2960	
2928	21	2	2879	2962	2894	-1	-1	2881	2961	
2929	21	2	2880	2881	2887	2964	2882	2962	-1	
2930	21	2	2964	-1	2886	2880	2952	2963	-1	
2931	21	2	-1	2884	2883	2965	2891	-1	2885	
2932	21	2	-1	-1	2963	2885	2966	2886	2965	
2933	21	2	2888	2894	2962	-1	-1	2887	2967	
2934	21	2	2967	2895	2889	-1	-1	-1	2888	
2935	21	2	-1	2892	-1	2966	2890	2948	2968	
2936	21	2	-1	2893	-1	2891	2968	2893	2890	
2937	21	2	-1	2968	-1	-1	2893	2969	2892	
2938	21	2	2895	2967	2961	2896	-1	-1	2894	
2939	22	2	2898	2897	2970	-1	-1	-1	2906	
2940	22	2	2900	2970	2897	-1	-1	-1	2905	
2941	22	2	2970	2900	2899	2971	-1	-1	2904	
2942	22	2	-1	-1	2971	2901	2972	-1	2907	
2943	22	2	-1	2903	-1	2972	2902	2973	2908	
2944	22	2	-1	2973	-1	-1	2973	2903	2974	
2945	22	2	2904	2905	2906	2975	-1	2912	2970	
2946	22	2	2975	-1	2907	2904	2976	2913	2971	
2947	22	2	-1	-1	2976	2908	2907	2977	2972	
2948	22	3	-1	2910	-1	2977	2909	2935	2978	2908(6)
2949	22	3	-1	2911	-1	2917	2978	2911	2909	

Para-meter	ℓ	d	Simples							Other
			1	2	3	4	5	6	7	
2950	22	3	-1	2978	-1	-1	2911	2974	2910	
2951	22	2	2913	-1	-1	2912	2979	2975	-1	
2952	22	3	2979	-1	2914	2919	2930	2980	-1	2913(5)
2953	22	3	-1	-1	2980	2915	2977	2914	2981	
2954	22	3	-1	2916	2921	2981	2917	-1	2915	
2955	22	3	2919	2922	2922	2979	2918	2982	-1	
2956	22	3	2982	2923	2920	2980	-1	2919	2983	
2957	22	3	2924	2924	2983	2921	-1	-1	2920	
2958	22	3	2923	2982	2982	-1	-1	2922	2984	
2959	22	3	2984	2983	2924	-1	-1	-1	2923	
2960	22	2	2985	2926	2925	2986	-1	-1	2927	
2961	22	2	2926	2985	2938	-1	-1	-1	2928	
2962	22	2	2927	2928	2933	2987	-1	2929	2985	
2963	22	2	2987	-1	2932	2927	2988	2930	2986	
2964	22	2	2930	-1	-1	2929	2979	2987	-1	
2965	22	2	-1	-1	2986	2931	2989	-1	2932	
2966	22	2	-1	-1	2988	2935	2932	2977	2989	
2967	22	2	2934	2938	2985	-1	-1	-1	2933	
2968	22	2	-1	2937	-1	2989	2936	2990	2935	
2969	22	2	-1	2990	-1	-1	2990	2937	2974	
2970	23	2	2941	2940	2939	2991	-1	-1	2945	
2971	23	2	2991	-1	2942	2941	2992	-1	2946	
2972	23	2	-1	-1	2992	2943	2942	2993	2947	
2973	23	2	-1	2944	-1	2993	2944	2943	2994	
2974	23	3	-1	2994	-1	-1	2994	2950	2969	2944(7)
2975	23	2	2946	-1	-1	2945	2995	2951	2991	
2976	23	2	2995	-1	2947	-1	2946	2996	2992	
2977	23	3	-1	-1	2996	2948	2953	2966	2997	2947(6)
2978	23	3	-1	2950	-1	2997	2949	2994	2948	
2979	23	3	2952	-1	-1	2955	2964	2998	-1	2951(5)
2980	23	3	2998	-1	2953	2956	2996	2952	2999	
2981	23	3	-1	-1	2999	2954	2997	-1	2953	
2982	23	3	2956	2958	2958	2998	-1	2955	3000	
2983	23	3	3000	2959	2957	2999	-1	-1	2956	
2984	23	3	2959	3000	3000	-1	-1	-1	2958	
2985	23	2	2960	2961	2967	3001	-1	-1	2962	
2986	23	2	3001	-1	2965	2960	3002	-1	2963	
2987	23	2	2963	-1	-1	2962	3003	2964	3001	
2988	23	2	3003	-1	2966	-1	2963	2996	3002	
2989	23	2	-1	-1	3002	2968	2965	3004	2966	
2990	23	2	-1	2969	-1	3004	2969	2968	2994	
2991	24	2	2971	-1	-1	2970	3005	-1	2975	
2992	24	2	3005	-1	2972	-1	2971	3006	2976	
2993	24	2	-1	-1	3006	2973	-1	2972	3007	
2994	24	3	-1	2974	-1	3007	2974	2978	2990	2973(7)
2995	24	2	2976	-1	-1	-1	2975	3008	3005	
2996	24	3	3008	-1	2977	-1	2980	2988	3009	2976(6)
2997	24	3	-1	-1	3009	2978	2981	3007	2977	
2998	24	3	2980	-1	-1	2982	3008	2979	3010	
2999	24	3	3010	-1	2981	2983	3009	-1	2980	
3000	24	3	2983	2984	2984	3010	-1	-1	2982	
3001	24	2	2986	-1	-1	2985	3011	-1	2987	
3002	24	2	3011	-1	2989	-1	2986	3012	2988	
3003	24	2	2988	-1	-1	-1	2987	3008	3011	
3004	24	2	-1	-1	3012	2990	-1	2989	3007	
3005	25	2	2992	-1	-1	-1	2991	3013	2995	
3006	25	2	3013	-1	2993	-1	-1	2992	3014	
3007	25	3	-1	-1	3014	2994	-1	2997	3004	2993(7)
3008	25	3	2996	-1	-1	-1	2998	3003	3015	2995(6)
3009	25	3	3015	-1	2997	-1	2999	3014	2996	
3010	25	3	2999	-1	-1	3000	3015	-1	2998	
3011	25	2	3002	-1	-1	-1	3001	3016	3003	
3012	25	2	3016	-1	3004	-1	-1	3002	3014	
3013	26	2	3006	-1	-1	-1	-1	3005	3017	
3014	26	3	3017	-1	3007	-1	-1	3009	3012	3006(7)
3015	26	3	3009	-1	-1	-1	3010	3017	3008	
3016	26	2	3012	-1	-1	-1	-1	3011	3017	
3017	27	3	3014	-1	-1	-1	-1	3015	3016	3013(7)

REFERENCES

1. B. Boe, D. Collingwood, *A multiplicity one theorem for holomorphically induced representations*, Math. Z. **192** (1986), 265–282.
2. ______, *Multiplicity free categories of highest weight representations*, Comm. Alg. **18** (1990), 947–1032.
3. L. Casian, *Graded characters of induced representations of real reductive Lie groups I*, J. Alg. **123** (1989), 289–326.
4. L. Casian, D. Collingwood, *The Kazhdan-Lusztig conjecture for generalized Verma modules*, Math. Z. **195** (1987), 581–600.
5. ______, *Weight filtrations for induced representations of real reductive Lie groups*, Adv. Math. **73** (1989), 79–146.
6. D. Collingwood, *Harish-Chandra modules with the unique embedding property*, Trans. Amer. Math. Soc. **281** (1984), 1–48.
7. ______, *Representations of rank one Lie groups*, Pitman, Boston, 1985.
8. ______, *The $\mathfrak{n}$-homology of Harish-Chandra modules: Generalizing a theorem of Kostant*, Math. Ann. **272** (1985), 161–187.
9. ______, *Orbits and characters associated to highest weight representations*, Proc. Amer. Math. Soc. (to appear).
10. D. Collingwood, R. Irving, B. Shelton, *Filtrations on generalized Verma modules for Hermitian symmetric pairs*, J. reine angew. Math. **383** (1988), 54–86.
11. T. Enright, B. Shelton, *Categories of highest weight modules: applications to classical Hermitian symmetric pairs*, Mem. Amer. Math. Soc. **367** (1987).
12. D. Garfinkle, *The annihilators of irreducible Harish-Chandra modules for $SU(p,q)$ and other type A_{n-1} groups*, Amer. J. Math. (to appear).
13. H. Hecht, *The characters of some representations of Harish-Chandra*, Math. Ann. **219** (1976), 213–226.
14. S. Helgason, *Differential geometry, Lie groups, and symmetric spaces*, Academic Press, New York, 1978.
15. R. Irving, *Projective modules in the category $\mathcal{O}_S$: Loewy series*, Trans. Amer. Math. Soc. **291** (1985), 733-754.
16. G. Lusztig, D. Vogan, *Singularities of closures of K-orbits on flag manifolds*, Inv. Math. **71** (1983), 365–379.
17. T. Matsuki, T. Oshima, *Embeddings of discrete series in principal series*, The orbit method in representation theory (Copenhagen, 1988), Progr. Math., vol. 82, Birkhäuser, Boston, 1990, pp. 147–175.
18. R. Richardson, T. Springer, *The Bruhat order on symmetric varieties*, Geometriae Dedicata **35** (1990), 389–436.

19. M. Sigiura, *Conjugate classes of Cartan subalgebras in real semisimple Lie algebras*, J. Math. Soc. Japan **11** (1959), 374–434.
20. B. Speh, D. Vogan, *Reducibility of generalized principal series representations*, Acta. Math. **145** (1980), 227–299.
21. D. Vogan, *Representations of real reductive Lie groups*, Birkhäuser, Boston, 1981.
22. ______, *The Kazhdan-Lusztig conjecture for real reductive groups*, in Representation theory of reductive groups (P. C. Trombi, ed.), Birkhäuser, Boston, 1983.
23. ______, *Irreducible characters of semisimple Lie groups III: proof of the Kazhdan-Lusztig conjectures in the integral case*, Inv. Math. **71** (1983), 381–417.
24. ______, *Irreducible characters of semisimple Lie groups IV: character multiplicity duality*, Duke Math. J. **49** (1982), 943–1073.
25. N. Wallach, *The asymptotic behavior of holomorphic representations*, Soc. Math. France **15** (1984), 291–305.

Department of Mathematics, University of Georgia, Athens, Georgia 30602
E-mail address: brian@joe.math.uga.edu

Department of Mathematics, University of Washington, Seattle, Washington 98195
E-mail address: colling@math.washington.edu

Editorial Information

To be published in the *Memoirs*, a paper must be correct, new, nontrivial, and significant. Further, it must be well written and of interest to a substantial number of mathematicians. Piecemeal results, such as an inconclusive step toward an unproved major theorem or a minor variation on a known result, are in general not acceptable for publication. *Transactions* Editors shall solicit and encourage publication of worthy papers. Papers appearing in *Memoirs* are generally longer than those appearing in *Transactions* with which it shares an editorial committee.

As of January 1, 1993, the backlog for this journal was approximately 7 volumes. This estimate is the result of dividing the number of manuscripts for this journal in the Providence office that have not yet gone to the printer on the above date by the average number of monographs per volume over the previous twelve months, reduced by the number of issues published in four months (the time necessary for preparing an issue for the printer). (There are 6 volumes per year, each containing at least 4 numbers.)

A Copyright Transfer Agreement is required before a paper will be published in this journal. By submitting a paper to this journal, authors certify that the manuscript has not been submitted to nor is it under consideration for publication by another journal, conference proceedings, or similar publication.

Information for Authors

Memoirs are printed by photo-offset from camera copy fully prepared by the author. This means that the finished book will look exactly like the copy submitted.

The paper must contain a *descriptive title* and an *abstract* that summarizes the article in language suitable for workers in the general field (algebra, analysis, etc.). The *descriptive title* should be short, but informative; useless or vague phrases such as "some remarks about" or "concerning" should be avoided. The *abstract* should be at least one complete sentence, and at most 300 words. Included with the footnotes to the paper, there should be the 1991 *Mathematics Subject Classification* representing the primary and secondary subjects of the article. This may be followed by a list of *key words and phrases* describing the subject matter of the article and taken from it. A list of the numbers may be found in the annual index of *Mathematical Reviews*, published with the December issue starting in 1990, as well as from the electronic service e-MATH [**telnet e-MATH.ams.org** (or **telnet 130.44.1.100**). Login and password are **e-math**]. For journal abbreviations used in bibliographies, see the list of serials in the latest *Mathematical Reviews* annual index. When the manuscript is submitted, authors should supply the editor with electronic addresses if available. These will be printed after the postal address at the end of each article.

Electronically-prepared manuscripts. The AMS encourages submission of electronically-prepared manuscripts in AMS-TEX or AMS-LATEX. To this end, the Society has prepared "preprint" style files, specifically the amsppt style of AMS-TEX and the amsart style of AMS-LATEX, which will simplify the work of authors and of the production staff. Those authors who make use of these style files from the beginning of the writing process will further reduce their own effort.

Guidelines for Preparing Electronic Manuscripts provide additional assistance and are available for use with either $\mathcal{A}_{\mathcal{M}}\mathcal{S}$-TEX or $\mathcal{A}_{\mathcal{M}}\mathcal{S}$-LATEX. Authors with FTP access may obtain these *Guidelines* from the Society's Internet node e-MATH.ams.org (130.44.1.100). For those without FTP access they can be obtained free of charge from the e-mail address guide-elec@math.ams.org (Internet) or from the Publications Department, P. O. Box 6248, Providence, RI 02940-6248. When requesting *Guidelines* please specify which version you want.

Electronic manuscripts should be sent to the Providence office only after the paper has been accepted for publication. Please send electronically prepared manuscript files via e-mail to pub-submit@math.ams.org (Internet) or on diskettes to the Publications Department address listed above. When submitting electronic manuscripts please be sure to include a message indicating in which publication the paper has been accepted.

For papers not prepared electronically, model paper may be obtained free of charge from the Editorial Department at the address below.

Two copies of the paper should be sent directly to the appropriate Editor and the author should keep one copy. At that time authors should indicate if the paper has been prepared using $\mathcal{A}_{\mathcal{M}}\mathcal{S}$-TEX or $\mathcal{A}_{\mathcal{M}}\mathcal{S}$-LATEX. The *Guide for Authors of Memoirs* gives detailed information on preparing papers for *Memoirs* and may be obtained free of charge from AMS, Editorial Department, P. O. Box 6248, Providence, RI 02940-6248. The *Manual for Authors of Mathematical Papers* should be consulted for symbols and style conventions. The *Manual* may be obtained free of charge from the e-mail address cust-serv@math.ams.org or from the Customer Services Department, at the address above.

Any inquiries concerning a paper that has been accepted for publication should be sent directly to the Editorial Department, American Mathematical Society, P. O. Box 6248, Providence, RI 02940-6248.

Recent Titles in This Series

(Continued from the front of this publication)

454 **Donald A. Dawson and Edwin A. Perkins,** Historical processes, 1991
453 **Alfred S. Cavaretta, Wolfgang Dahmen, and Charles A. Micchelli,** Stationary subdivision, 1991
452 **Brian S. Thomson,** Derivates of interval functions, 1991
451 **Rolf Schön,** Effective algebraic topology, 1991
450 **Ernst Dieterich,** Solution of a non-domestic tame classification problem from integral representation theory of finite groups ($\Lambda = RC_3, v(3) = 4$), 1991
449 **Michael Slack,** A classification theorem for homotopy commutative H-spaces with finitely generated mod 2 cohomology rings, 1991
448 **Norman Levenberg and Hiroshi Yamaguchi,** The metric induced by the Robin function, 1991
447 **Joseph Zaks,** No nine neighborly tetrahedra exist, 1991
446 **Gary R. Lawlor,** A sufficient criterion for a cone to be area-minimizing, 1991
445 **S. Argyros, M. Lambrou, and W. E. Longstaff,** Atomic Boolean subspace lattices and applications to the theory of bases, 1991
444 **Haruo Tsukada,** String path integral realization of vertex operator algebras, 1991
443 **D. J. Benson and F. R. Cohen,** Mapping class groups of low genus and their cohomology, 1991
442 **Rodolfo H. Torres,** Boundedness results for operators with singular kernels on distribution spaces, 1991
441 **Gary M. Seitz,** Maximal subgroups of exceptional algebraic groups, 1991
440 **Bjorn Jawerth and Mario Milman,** Extrapolation theory with applications, 1991
439 **Brian Parshall and Jian-pan Wang,** Quantum linear groups, 1991
438 **Angelo Felice Lopez,** Noether-Lefschetz theory and the Picard group of projective surfaces, 1991
437 **Dennis A. Hejhal,** Regular *b*-groups, degenerating Riemann surfaces, and spectral theory, 1990
436 **J. E. Marsden, R. Montgomery, and T. Ratiu,** Reduction, symmetry, and phase mechanics, 1990
435 **Aloys Krieg,** Hecke algebras, 1990
434 **François Treves,** Homotopy formulas in the tangential Cauchy-Riemann complex, 1990
433 **Boris Youssin,** Newton polyhedra without coordinates Newton polyhedra of ideals, 1990
432 **M. W. Liebeck, C. E. Praeger, and J. Saxl,** The maximal factorizations of the finite simple groups and their automorphism groups, 1990
431 **Thomas G. Goodwillie,** A multiple disjunction lemma for smooth concordance embeddings, 1990
430 **G. M. Benkart, D. J. Britten, and F. W. Lemire,** Stability in modules for classical Lie algebras: A constructive approach, 1990
429 **Etsuko Bannai,** Positive definite unimodular lattices with trivial automorphism groups, 1990
428 **Loren N. Argabright and Jesús Gil de Lamadrid,** Almost periodic measures, 1990
427 **Tim D. Cochran,** Derivatives of links: Milnor's concordance invariants and Massey's products, 1990
426 **Jan Mycielski, Pavel Pudlák, and Alan S. Stern,** A lattice of chapters of mathematics: Interpretations between theorems, 1990
425 **Wiesław Pawłucki,** Points de Nash des ensembles sous-analytiques, 1990

(See the AMS catalog for earlier titles)